WITH

D0404159

"Brendan Kelly uses his expertise in Chinese .... to draw a connection between the environmental excesses that have led us to the age of climate change and the individual excesses that lead to depletion, imbalance, and disease. His fascinating book offers fresh, clear insight into the root causes of both, as well as a map toward greater personal and environmental health and balance."

—Kristin Kimball, author of *The Dirty Life*;
farmer and co-founder of Essex Farm

"The human body is a mirror of our home planet. *The Yin and Yang of Climate Crisis* not only teaches that the body and earth are connected, it explains the need for healing both."

—Scott Frazier, member of the Crow/Santee tribe;
founder and director of Project Indigenous

"Brendan Kelly invites us to set aside our typical ways of thinking—reductionistic, fragmented, and dualistic—and to embrace a more holistic and systemic mode of thought. Using the conceptual framework of Chinese medicine, Kelly incisively diagnoses the source of what most ails us, both personally and collectively, and offers guidelines for meaningful change. What could be more important?"

—John Christopher, PhD, Fellow of the American Psychological
Association and the Mind & Life Institutes

"With a radically expansive understanding of holistic principles, Brendan Kelly shows us that the remedy for our individual bodies and our planet-body is the same. He invites us to look at the deeper causes of our situation and, in so doing, to experience deeper purpose and connection with life. This book is a must-read for anyone who cares about the earth."

— Sarah Von Hoy, PhD, LAc, professor at Goddard College

"*The Yin and Yang of Climate Crisis* is a brilliantly written, far-sighted exploration of climate change using Chinese medicine as an ancient way to see personal mind-body inflammation as deeply interrelated with the earth's rising heat. This book empowers people to embrace cooler lifestyles and less inflammatory diets to help rebalance their own lives and in a small, yet organically powerful, way."

—Susan Green, PhD, department chair of Behavioral Sciences and director of Wellness & Alternative Medicine, Johnson State College

The Yin and Yang of Climate Crisis

Healing Personal, Cultural, and Ecological Imbalance with Chinese Medicine

BRENDAN KELLY, LAc

North Atlantic Books
Berkeley, California

Published by
North Atlantic Books
Berkeley, California

Cover image © shutterstock.com/Alexandr79
Cover and book design by Mary Ann Casler
Printed in the United States of America

The Yin and Yang of Climate Crisis: Healing Personal, Cultural, and Ecological Imbalance with Chinese Medicine is sponsored and published by the Society for the Study of Native Arts and Sciences (dba North Atlantic Books), an educational nonprofit based in Berkeley, California, that collaborates with partners to develop cross-cultural perspectives, nurture holistic views of art, science, the humanities, and healing, and seed personal and global transformation by publishing work on the relationship of body, spirit, and nature.

North Atlantic Books' publications are available through most bookstores. For further information, visit our website at www.northatlanticbooks.com or call 800-733-3000.

Library of Congress Cataloging-in-Publication Data
Kelly, Brendan (Brendan D.), author.
  The yin and yang of climate crisis : healing personal, cultural, and ecological imbalance with Chinese medicine / Brendan Kelly.
       pages cm
Summary: "Examines the current climate crisis through the lens of Chinese medicine."—Provided by publisher.
ISBN 978-1-58394-951-1 (paperback) — ISBN 978-1-58394-952-8 (e-book)
1. Climatic changes—Health aspects. 2. Medical climatology. 3. Human beings—Effect of climate on. 4. Medicine, Chinese. I. Title.
RA793.K45 2015
616.9'88—dc23   2015001171

1 2 3 4 5 6 7 8 9 United 20 19 18 17 16 15

Printed on recycled paper

Dedicated to my beautiful wife, Liz, who helped make this book happen in innumerable ways, both large and small

Also dedicated to my parents, Joan and Richard, for a lifetime of love, support, and encouragement

Acknowledgments

To my wife, Liz—thank you for several years of keeping me fed, which provided me days at a time to write and research. And thank you for a decade of listening to me talk about the ideas that would become this book.

To the patients at our clinic—thank you for the opportunity to see the depth of healing possible with Chinese medicine.

To my current teachers of Chinese medicine, especially Wolfe Lowenthal and Jeffrey Yuen—thank you for all that you've given me and all of your students.

To the teachers of my teachers of Chinese medicine, including Huang Di, Sun Si Miao, Zhang Zhong-jing, and Cheng Man-ch'ing—thank you for the wisdom you've imparted to generations of students and practitioners.

And to Tim McKee, Richard Grossinger, and everyone at North Atlantic Books—thank you for seeing the potential in a Chinese medicine understanding of climate change.

Contents

Introduction   xi

Chapter 1: The Sickness of Climate Change   1

Chapter 2: The Science of Climate Change:
    A Chinese Medicine Perspective   15

Chapter 3: The Meaning of Climate Change:
    The Personal Level   35

Chapter 4: Feeding the Fire of Climate Change   49

Chapter 5: The Water Phase: Oil and Jing   67

Chapter 6: The Consequences of Continuous Growth:
    Wood and Wind   81

Chapter 7: Quality Controls Quantity: Metal Controls Wood   109

Chapter 8: The Dissatisfaction of Too Much Fire   125

Chapter 9: Cancer and Climate Change:
  A Tale of Two Diagnoses   143

Chapter 10: Internal Climate Change: Tongue Diagnosis   167

Chapter 11: The Opportunities of Climate Change   187

Notes   199

Bibliography   217

Index   223

About the Author   229

Introduction

Often we're like fish in water. Because it's continuously all around them, there are stories about how fish aren't able to recognize the environment in which they live. For us, rather than not seeing the water all around us, we're often unable to see the assumptions that shape our lives.

How we see the world affects everything. It influences what we do, what we value, and how we define a good life. It is also the basis of our cultural institutions, including our economy, our medical system, and modern sciences like biology. Like fish in water, we're swimming in a sea of assumptions that are everywhere and affect all parts of our lives. Because they are all around us and permeate throughout our culture, we often accept without question our shared beliefs.

As you'll read throughout *The Yin and Yang of Climate Crisis,* it is these assumptions and our views of the world that are the deeper causes of our rapidly warming planet. Virtually everything we hear about climate change is from our usual, Western perspective. Most of the discussion about the crisis of global warming focuses on external issues: calls to reduce greenhouse gas emissions, increase carbon sequestration, buy and eat locally, and challenge continuous economic growth. These remedies are undoubtedly important, but if we were to look at climate change from a different vantage point, we can see how what is happening in the environment around us is also happening within us. In particular, we can understand that the severity of climate change speaks to deeper and more wide-reaching philosophical and spiritual issues.

The essential importance of stepping outside our usual view of the world to look at the climate crisis is that the transformation we now need requires us to see clearly the consequences of our personal and cultural beliefs. Maintaining the usual perspectives about what signifies a life worth living, how we view nature, and how we treat sickness will continue to lead us down the same path we're on. As we'll discuss with the chapters to follow, this path has not only led us to a place of dramatic climate destabilization, it has also had similar, deep-reaching effects on all aspects of our lives.

For thousands of years, Chinese medicine has understood the world holistically. From this Eastern perspective, all of our organs are interconnected, and the physical, mental, and emotional aspects of our lives are linked together. Developed over millennia, Chinese medicine recognizes that this holism within us is a reflection of interdependence in the world around us. Not only are we connected to other people and to our culture, we are inseparably connected to nature.

The chapters that follow blend the external focus of environmentalism—Western science, policy issues, and regulations—with the internal focus of Chinese medicine—personal health, balancing Qi, and diet. From this new, combined perspective, climate change and its literal realities, such as melting ice caps, dying forests, and floods, can be understood as a symptom of deeper issues—both within us as individuals and within our country and culture.

When treating conditions like headaches and back pain, acupuncturists and Chinese herbalists attempt to understand the source of the symptom rather than solely addressing the symptom itself. In the treatment room, it is important to look below the surface to diagnose and treat root causes because symptoms tell us that something is out of balance. To use the nature-based language common in Chinese medicine, symptoms are the branches that extend to the physical, surface level; the underlying causes, however, are internal, stemming from roots deeper below the surface.

The Yin and Yang of Climate Crisis unfolds Chinese medicine's stance on symptoms: they are messengers trying to get our attention. Symptoms of all kinds and severity—within us or the climate—communicate that something is out of balance. The more severe the symptom, the more urgent the message. In my ten years of clinical work as an acupuncturist and herbalist, I've found that it's common to see patients who have multiple symptoms. Rather than look at each of these issues in isolation, Chinese medicine treats symptoms as interconnected. Applying this diagnostic perspective to our planet, the chapters that follow discuss climate change as a form of sickness and offer several interconnected, deep-reaching remedies.

The big-picture perspective that Chinese medicine offers readily lends itself to not only addressing issues with our own personal health but also the condition of our culture. A crucial aspect to the holism of Chinese medicine is that the big picture and the little picture are very similar. In other words, what's happening within us individually is a reflection of what's happening in our culture, and vice versa.

Extending this view to the scale of the planet, it's clear that Chinese medicine has much to offer to the discourse surrounding our rapidly warming and destabilizing planet. As has been written about extensively, the climate is warming due to greenhouse gases and burning fossil fuels. Given its well-developed and insightful understanding of sickness and root causes, an Eastern perspective on the Western science of climate change helps us see patterns in the vast amount of global climate data.

In the treatment room, a practitioner of Chinese medicine looks for patterns and connections of different symptoms and diagnoses. Similarly, when we look at the data on climate science through the lens of Chinese medicine, we can see a clear pattern of what is happening to the planet. In particular, the planet is warming rapidly as its ability to maintain a cool and stable climate is decreasing.

Chinese medicine's ability to see clear connections that might otherwise appear separate comes from being a medicinal tradition based on nature. As we'll discuss throughout the rest of the book, major aspects of diagnostic and treatment traditions in Chinese medicine are based on the seasons and weather. The Five Elements tradition, also known as the Five Phases, provides an extremely well developed understanding of how the effects of the seasons in nature are mirrored by similar effects within us. Part of this tradition includes how each of the phases correspond to different seasons and how health is maintained from both increasing and decreasing these seasonal effects. In addition, the condition of our internal climate—the balance of hot, cold, damp, and dry—also moves us toward health or sickness.

Chinese medicine also helps us see that the root causes of climate change are affecting our lives in other ways we may have not considered. On a personal level, this manifests in the things we choose to value and how we communicate with one another. Along with these mental and emotional issues is the reality of our physical health. The very significant rates of cancer in our country and the number of people who are estimated to die from the condition—both of which are projected to increase yearly—also indicate how the serious imbalances of the climate are mirrored in our own internal environment.

When viewed through Chinese medicine's holistic perspective, it's not surprising that the state of the climate—severe storms, floods, and droughts globally—and our internal condition—the projection of over 120 million cancer diagnoses—is also being mirrored in our country's culture. How we've structured our economy, how we practice and receive medicine, and how we view nature are all a reflection of these deeper issues.

By examining and addressing these root issues, we have the opportunity to treat the underlying causes of climate change. In doing so, we have the opportunity to heal the people and the institutions that have most contributed to the condition, namely those in the United States. Specific case studies from my clinical work are presented to make clear the connection between the condition of the individual, the country, and the climate.

In chapter 1, we'll look at the recent severe storms in Vermont—where I live and practice Chinese medicine—as one example of what's happening globally. We'll also look at one common Chinese diagnosis as a small-scale example of what's happening in the climate and in our country. Chapter 1 gives a brief background of Chinese medicine's development over thousands of years, especially its understanding of Yin and Yang and use of inductive thinking. Finally, the chapter discusses how the imbalances between the cooling and warming aspects of the climate also appear in our own lives.

In chapter 2, we'll dive into the Western science of climate change, including some of its history. The studies and statistics in this chapter are necessary to investigate the patterns found in Western science from a Chinese medicinal view. Through this Eastern perspective of Western science, a clear pattern emerges of what's happening to the climate.

Chapter 3 looks deeper at the reflection of imbalance within us, both individually and collectively. We'll talk about how consumerism, our values, and our stories all speak to the underlying causes of our warming planet. The chapter concludes with suggestions for how we can internally address the causes of climate change.

In chapter 4, we'll discuss how the amount of heat we create internally through our overconsumption of coffee reflects the rapidly overheating climate. We'll also look at the long list of physical, mental, and emotional symptoms that result from drinking coffee, as well as cooling substitutes.

Chapter 5 begins the discussion of the Five Phases tradition of Chinese medicine—this chapter focuses on Water, in particular. This chapter introduces the Chinese medicine concept of *jing*, which are our deep, concentrated reserves of energy passed along to us by our parents. Roughly analogous to the Western idea of DNA, jing is what makes us the unique individuals we are. This chapter looks at oil as the jing of the planet, as well as how continuing to burn oil contributes to the lack of generation foresight we urgently need to address the root causes of climate change.

Chapter 6 discusses the next phase—Wood—and how our belief in continuous growth speaks to another root cause of climate change. In particular, we'll examine how basing our economy on the idea that it can grow forever not only has severe ecological consequences but also promotes an excess of Wood within us and our country. This chapter also presents the Chinese medicine concept of wind, which is what creates change and movement within us, and the concept of the Sheng cycle, also called the Nourishing cycle. The Sheng cycle describes how different parts of nature feed and support other parts, and how this same nourishing dynamic appears within us as well.

In chapter 7, we discuss the Metal phase, which includes a sense of deep meaning and an experience of something greater than our individual lives. In Chinese medicine, this includes an experience of the sacred in all aspects of our lives, including what we value. In this chapter, the Chinese medicine understanding of the K'o cycle, also called the Control cycle, is introduced. Unlike the Sheng cycle, the K'o cycle limits and controls.

In chapter 8, we'll discuss the Fire phase, burning too bright and too hot. Mirroring the heat in the climate is an excess of fire within us, exemplified by our dissatisfaction with how we communicate with each other. In particular, the amount of emails, texts, and tweets we send and receive each day often creates too much stimulation and too little satisfaction. A lack of satisfaction means not only a lack of meaningful connection with the people and the world around us but also a lack of understanding of who we are. Finally, we discuss how the excess of the Wood contributes to the issues of the Fire, as seen in the Sheng or Nourishing cycle relationship between the two phases.

Chapter 9 makes the link between the extraordinary rates of cancer in the United States and climate change. Looking at both conditions through the lens of Chinese medicine, a clear connection appears— namely, the diagnosis of heat. Taking a big-picture view of cancer, we look at how our quest for continuous economic growth is reflected in the growth of unhealthy cells within us. We'll also discuss how the Western approach of waiting for a problem to occur before we attempt to address it—both with our rapidly warming planet and the overgrowth of unhealthy, cancerous cells—speaks to our cultural imbalances.

Next, Chinese medicine's use of tongue diagnosis as a tool to understand our internal condition is introduced. Chapter 10 provides the well-established logic behind tongue diagnosis and what the size, shape, color, and coating of the tongue says about our internal condition. You'll be encouraged to look at your own tongue to understand more clearly how the global climate is a reflection of our internal condition. In particular, tongue diagnosis makes it clear that the warming climate is similar to the warming of our internal environment.

The last chapter, chapter 11, reimagines the potential catastrophe of climate change as an opportunity. Part of the remedy for climate change is slowing down, which naturally occurs during the harvest— the Earth phase. Here, we see how Chinese medicine helps us recognize that increasing temperatures and melting glaciers can be good news. According to Chinese medicine, when something realizes its full expression, it transforms into something else. It doesn't require a belief in change to understand that the underlying causes of climate change are, in fact, leading us toward sustainability. The chapter concludes that what we need to do now is to allow these changes to occur and not attempt to stand in the way.

Rather than being separate conditions, our health, the health of Western culture, and the well-being of the planet involve similar issues happening on different scales. The holism embedded in Chinese medicine allows us to see that the dramatic changes we're experiencing in the climate are mirrored in the imbalances of our own individual lives and in the United States. What an Eastern view of the climate crisis offers is the opportunity to understand the root causes of our rapidly warming planet. Rather than being like fish that can't see the water all around them, in applying Chinese medicine to climate change, we have the opportunity to examine clearly the imbalances arising from how we view the world. This is good medicine not only for the well-being of the climate but also a potent remedy to help lead us back to health.

CHAPTER 1

The Sickness of Climate Change

In August 28 of 2011, northern Vermont received seven inches of rain. Tropical Storm Irene alternated between a steady and torrential rain for a full twenty-four hours. Our office—where we practice acupuncture and herbal medicine—overlooks the Winooski River in the town of Burlington. In one day, the character of the river had changed dramatically. On the Saturday before the storm, you could walk across the 150-foot expanse between its banks. The water was shallow and flat, quietly flowing past large and small exposed rocks.

After a day of continuous rain on Sunday, by Monday morning there was a very different view from our clinic. The river was raging and had become a Class V rapid outside our office's front door. Large rocks that previously sat well above the waterline were now shooting water ten feet straight up into the air. There were twelve- to fifteen-foot swells where the river ascended and cascaded down boulders and other obstacles in its path. The quiet from two days earlier had been replaced with the raw thundering of the powerfully rushing water. It was so loud that the several dozen people standing on the embankment above the river were scarcely talking to each other.

There were likely two reasons for how quiet they were. First, the river was so loud that it was nearly impossible to communicate. Second, we were still in the midst of digging out from the effects of the previous flood that occurred just five months earlier. Those of us standing above the river were likely aware that if it continued to rise, there was nothing we could do about it.

What was before the storm a five- to six-foot embankment of rock and dirt at the edge of our office parking lot was now covered in frothy, mud-colored water. The river started to lap over the top of the three-foot-high cement reinforcement blocks and spread throughout the parking lot. The building's maintenance workers began to pile sand bags along the river's edge and run a sump pump at the low spot of the parking lot to try to move the pooling water. But it was clear that if the river continued to rise, it wouldn't make much difference. We knew there was a strong possibility of another major flood. During my five hours at the office, the river had risen as many feet.

What happened in Burlington also happened in much of the rest of Vermont, and is part of a larger, global process. Across the state, whole bridges and roads had been washed away. In the town of Brattleboro, located in the southeast corner of the state, the entire downtown had been submerged under several feet of water as multiple rivers nearby breeched their banks. In Rockingham, north of Brattleboro, a covered bridge originally built in the early 1800s had washed downstream, and a video of the incident was posted on YouTube. In the middle of the state, the downtown area of Waterbury, including much of Main Street, was under several feet of water. Speaking on what had been happening across the state, National Weather Service hydrologist Greg Hanson said early Monday, "From what we're seeing, this is one of the top weather-related disasters in Vermont's history."[1]

As my wife, Liz, and I periodically looked down from our third-floor office window, what we were seeing directly below our Chinese medicine clinic felt like an important analogy for what was happening all around Vermont. Acupuncture needles, meridian charts, and medical texts surrounded us inside the office, and outside the water was surging and rising below us. The river that was low, smooth, and quiet just thirty-six hours before was now raging, gushing high into the air, and on the verge of flooding.

With this view of the river from our clinic in mind, the importance of Chinese medicine's emphasis on looking below the surface becomes clear. Rather than focusing on treating only symptoms themselves, for thousands of years Chinese medicine has stressed the importance of understanding where symptoms come from. Even though the flooding severely affected the people and the landscape of Vermont, it was a sign—a symptom—of larger and deeper issues.

Chinese Medicine's View of Symptoms: From Headaches to Flooding Rivers

When treating patients, Chinese medicine practitioners work to treat the root causes of symptoms, instead of only the symptoms themselves. Whether mild or life-threatening, symptoms are indicators of deeper imbalances. With headaches, for example, there are many methods for relieving and eliminating the associated pain. But if we don't attempt to understand the root causes of the headache, we aren't listening to what the symptom is trying to tell us. Rather than seeing them as conditions to eliminate, we can approach symptoms as messengers and listen to the messages they are delivering. Symptoms of all kinds—from headaches to flooding rivers—are trying to get our attention and let us know that something is out of balance. The more severe a symptom, the more urgently the messenger is trying to deliver the message.

There are, of course, many ways to greet the messenger. We can ignore it, and hope that our symptom will just go away. Sometimes this seems to work well, especially if the symptoms are mild and we are generally healthy. In many cases, however, the same messenger—or a different one—will return because we haven't taken the time to listen to or translate the message.

Part of the difficulty in listening to symptoms is that our dominant cultural and medical view espouses an approach based on separation.

For example, we assume that we are separate from nature, our organs are separate from our other organs, and our body is separate from our mind and emotions. This separation makes it easy to see a headache as just that—pain in the head. However, if we recognize that there is an inherent interconnection, both within us and within the wider world, we then have the opportunity to understand and address the deeper causes of symptoms.

Just as an acupuncturist or Chinese herbalist asks questions when seeking to treat the causes of a headache, there are essential questions that needed to be asked about what caused the river below our office to start to flood the parking lot. As is well documented, climate change is leading to an increase in storms, floods, and other severe weather around the world. These are not theoretical issues that might happen at some point in the future—the realities of climate change are happening here and now.

But instead of merely looking at the effects of climate change, the perspective of Chinese medicine asks us to try to understand where this change is coming from. If we think of severe weather as a symptom, we can then ask a question of real importance: *What is the symptom of climate change trying to tell us?*

The typical approach, however, to our increasingly dramatic weather is to examine the science of what is occurring and the public policy issues related to the crisis. We often hear about the increase in greenhouse gas emissions, the subsidization of the fossil fuel industry, and the corporate greed that impacts environmental policies. Understanding the science and policies of climate change is undoubtedly important. Though the policies are very real and very significant, they are also symptoms of deeper issues. Our assumptions are some of the deeper causes of the climate crisis—things like how we see the world, what we value, how we define success, and the stories we choose to tell and listen to.

Additionally, the science and policies are based on the assumption that climate change is happening outside us. The research is clear that we are living on a planet that is warming and destabilizing, resulting in more torrential rains and floods like what we experienced in Vermont. Looking at what is happening through the lens of Chinese medicine, however, we can also see that what is occurring in the climate is occurring within us. In other words, the warming and destabilization of the climate mirrors our own personal imbalances, as well as the imbalances in our country and our culture. Just as the planet is rapidly warming and destabilizing, we as individuals are dramatically and unsustainably overstimulated. And just as the climate's ability to hold greenhouse gases has been compromised, our own internal peace has also been compromised. Here in the United States, our individual overemphasis on newness, having more, and doing more is a reflection of what is happening with the climate on a global scale.

All symptoms potentially have a cure. In the face of the floods in Vermont and similar severe weather globally that now confronts us, an important question arises: *What is the remedy for the sickness of climate change?* But before we can begin to answer that question, we need to understand what is causing the sickness itself. Symptoms of all kinds and on all scales appear because things are out of balance. Even a symptom that is as severe and wide-reaching as climate change has a set of remedies that can address it through treating its underlying causes. If we do take this approach, as happens with individuals who receive acupuncture treatments and take herbal medicine, not only is it possible for the symptom of climate change itself to be treated, but a deep sense of well-being can emerge within the country and the culture that is responsible for having created the crisis.

The Little Picture and the Big Picture: The Microcosm and Macrocosm

A middle-aged woman arrived at our clinic after calling for an appointment earlier that week. I asked what had brought her in, and she began to describe symptoms that have become synonymous in America with women and aging—hot flashes that wake her at night. She also described a generally shallow and restless sleep, persistent low-grade anxiety, high-pitched ringing in her ears, and occasional severe, debilitating headaches.

After about forty-five minutes of questions about her life and health, I requested that she lie face-up on the treatment table. Standing next to the table and facing her, I took her left hand in my left hand, palms facing each other. I positioned my right index finger just below where her wrist bends, on the thumb side, and then my right middle finger and ring finger below that. I repeated the process on her right side. My fingers were placed just above where blood flows in her arteries, and I focused my attention for a few minutes on each side, applying light pressure. While doing so, I listened for information even more subtle than the number of heartbeats per minute, namely the condition of what Chinese medicine calls Qi. Rather than listen to the pulse of the blood, I listened to the Qi's pulses of the twelve organs, six on each wrist.

Chinese medicine's long history—which has only recently been introduced to the West—has produced many different translations of the meaning of Qi. A scholarly and literal translation is "the finest influence of matter,"[2] while a more general understanding is "Qi governs the shape and activity of the body and its process of forming and organizing itself."[3] Regardless of the particular definition used, pulse diagnoses can tell us the condition of a person's health and their internal organs by listening to the amount and the quality of Qi that is moving along the surface of the wrist at different locations.

After listening to the different pulses, I asked Jenny (whose name I have changed to protect her identity) to stick out her tongue several times. Unlike in Western medical clinics, where patients open their mouths so doctors can examine the back of their throats, in Chinese medicine we can understand the condition of the internal organs based on the condition of the tongue. As we'll discuss later, the size, shape and color of our tongue tells us what's happening internally.

Through several thousand years of continuous refinement, Chinese medicine has developed diagnostic methods that enable the practitioner to understand what is happening inside the body by assessing what is happening externally. Based on several millennia of understanding is the recognition that the small picture—for example, the microcosm of Jenny's tongue and the Qi in one part of her wrist—is a reflection of the big picture—her overall macrocosm of health. These diagnostic methods are tools that allow us to deeply understand people's internal conditions as well as a concrete example of Chinese medicine's larger philosophical approach. If we view the world and our lives through the lens of separation, Chinese medical diagnosis might seem like an outdated superstition, giving rise to the skeptic's question: *How can placing your fingers on someone's wrist and looking at their tongue tell you anything about what is happening in the body?*

The answer comes from understanding the interconnection that is embedded within Chinese medicine. Rather than seeing the world through a lens of isolation and separation, as does Western medicine, Chinese medicine views the world with a perspective of connection. From this holistic vantage point, the nature of something can be seen if we look at it in great detail or from the panoramic view. Some of us in the West have come to believe that we are individual islands unto ourselves—that we are separate not only from the people around us but from the world in general. Similarly, we might view our organs as separate entities and our body as separate from our mind and emotions. But the older—and in my view, more developed and insightful—Chinese medicinal perspective is based on a different set of assumptions: one central understanding is that everything is indeed connected.

Equally important, this timeworn medical tradition is comfortable with inductive thinking, where we look for patterns and connections rather than the more common Western perspective that looks for absolute truth. Chinese medicine assumes that the microcosm and macrocosm reflect the same conditions and tendencies, with the only significant difference being scale. This is similar to the modern concept of the hologram, in which we can look at the whole image and see the picture clearly, or focus on a small piece of the image and again see the whole picture. In other words, the small picture and the big picture are, in fact, the same picture. The only real difference is one of scale.

The historical texts that serve as the basis for the study and practice of Chinese medicine have this assumption of interconnection and this holistic perspective woven within them. While some of Western medicine is now trying to include this interconnectedness and bigger picture view, Chinese medicine has already had this infused into its thinking, writing, and practice for several thousand years. For example, the *Nei Jing*, or *Yellow Emperor's Classic of Internal Medicine*, written several thousand years ago, describes how weather can promote health or sickness.[4] Though not often stated explicitly because it is assumed throughout the text, there is also an inherent understanding that we are not only connected to nature in general but also to the seasons. The *Yellow Emperor's Classic* states clearly that we as humans are a reflection of the universe as a whole. One translation of the text reads, a "human being is a small universe as the human body has everything that the universe has."[5]

Two other seminal Chinese medicine texts that are still prominently used in teaching and practice have a similar understanding of our connection to the natural world. The *Shang Han Lun*, published approximately 1,800 years ago, presents the viewpoint of the "School of Cold." In this tradition, cold weather can affect our health if our immunity is weakened, allowing cold to move deeper into the body and cause numerous internal conditions.[6] From a more Western biomedical perspective, the *Shang Han Lun* describes the progression of viral infections and their corresponding diagnostic and treatment principles.

The *Wen Bing Xue* provides the perspective of the "School of Heat," which states that all sickness begins with excess warmth. Again, from a more modern medical view, this text describes the progression of bacterial infections and associated diagnosis and treatment.[7] According to the *Wen Bing Xue,* heat can originate from environmental influences, and if not addressed, can travel deeper into the body and cause long-term chronic conditions. Both the School of Cold and the School of Heat understand that we are connected to the world around us and that what occurs in nature affects us, steering us toward health or sickness.

In addition to these and other foundational texts of Chinese medicine that articulate a perspective of interconnection, the practice of the medicine itself also speaks to its holographic understanding. In recognizing that the big picture and small picture are mirror images of each other, Chinese medical practitioners have used pulse and tongue diagnosis for millennia. It makes sense that we are able to understand the condition of our internal organs by placing our fingers lightly on the skin because the small picture—the Qi as it rises to the surface of the wrist—is a reflection of the larger picture—the condition of the internal organs. The same is true for understanding the condition of our organs by looking at the tongue, as it represents externally what is happening internally. In particular, the tip of the tongue corresponds to the organs in the chest, the middle of the tongue to the organs in the abdomen and the back of the tongue to the organs in the lower torso.

Similar to the relationship between the tongue and the internal organs, acupuncturists can insert small, extremely thin pin-like needles shallowly below the surface of the skin not only to affect the organs located much deeper in the body, but also to treat the mind and emotions. Even the Chinese medicinal conception of "organ" has holistic connotations. When Chinese medicine speaks about particular organs, they have both physical functions—often similar to more modern Western understandings—as well as emotional and spiritual characteristics.

For instance, in the *Yellow Emperor's Classic,* the Lungs are responsible for taking in oxygen on a physical level, as we understand in Western medicine, and they are also connected to the emotion of grief and our connection to the divine. As in Western medicine, the Liver is understood to filter the blood but also to be related to the expression of anger and our bigger vision of who we are. In addition to pumping blood, the Heart is also about the emotions of joy and love. It's also understood to be responsible for expressing who we are in the world, similar to the Western idea of it being important to be true to your heart.[8] When Chinese medicine refers to different organs, both their physical function and their mental and spiritual aspects are addressed.

Looking back at Jenny's condition, it was clear that both her pulses and her tongue were telling a similar story. Her internal condition was similar to that of many of us in the West—she had too much heat and not enough coolant. While this diagnosis can have numerous causes, it was clear from speaking to Jenny that a major part of her condition was from being too busy for too long. In her well-intended effort to be a productive employee, caring community member, loving mother and spouse, and to exercise regularly, Jenny had simply been too active physically and mentally for too many years. Whether we are taking a walk, talking to a coworker, or reading a book, we are using our Qi to do these things. This energy by its nature is warming, and when it has been overused for an extended period of time, it creates what Chinese medicine calls heat.

Heat is an excess of warmth and a state of overstimulation, which can eventually cause our internal fluids, or coolant, to evaporate, as when we leave a pot of water boiling on the stove. An excess of warmth and reduced internal fluids are called Yin-deficient heat, where the internal temperature not only increases but also the internal coolant decreases, like evaporating water. This creates a cycle where a loss of coolant increases the heat, which causes more coolant to evaporate, further increasing the heat.[9]

Yin and Yang, in Us and in the Climate

The concept of Yin and Yang are among the oldest and most developed existing medical understandings. Yin in nature corresponds with cold temperatures, the winter season, moisture such as rain and humidity, and night. Yang corresponds to warmth, summer, dryness, and day. Yin is also associated with water, and Yang with fire. As in the Yin-Yang circle, Yin is represented by black and Yang is represented by white. Similar to how they are seen in nature, Yin and Yang are fundamental influences within us—Yin is stillness, rest, and inactivity; Yang is movement, doing, and activity.[10]

Jenny's tongue and pulse indicated an excess of heat and a lack of coolant, particularly within her Kidney. Her tongue was red all over, and had a deep crack from front to back. The redness can be understood as the type of heat similar to a fire that is burning too brightly. The crack appears from Jenny drying out internally, as would soil exposed to a baking summer sun without any rain. The quality of Jenny's Qi was also dried out and hot. Her pulse was thin, indicating that her fluids were no longer full and were instead evaporating, and there was also a quality of excess force. Rather than being strong, the pulse felt as if it were agitated, pushing up quickly and abruptly against her skin. This indicates excess heat and feels as if the energy is trying to push its way up and out.

Jenny's individual condition bears a striking similarity to what climatologists are describing is happening to the planet. As has been exhaustively documented by the United Nations' Intergovernmental Panel on Climate Change (IPCC) and widely discussed in several popular books, our planet is rapidly getting warmer. In 2007, the IPCC issued its exhaustive fourth assessment report, summarizing published Western scientific data on the condition of the planet's climate. Along with Al Gore, it was awarded the Nobel Peace Prize in December of that year for assembling and analyzing the work of more than 2,500 scientific expert reviewers, with over 800 contributing authors and more than 450 lead authors from over 130 countries. To summarize the report's findings, there is a nearly universal scientific consensus that the planet is warming significantly and rapidly as a result of human actions. It is likely this will continue to contribute to more violent storms, dramatic melting of glaciers, increased floods and droughts, and general disruption of climate stability.

The report clearly states at the beginning of its summary that "warming of the climate system is unequivocal, as is now evident from observations of increase in global average air and ocean temperatures, widespread melting of snow and ice, and rising global average sea level."[11]

The vast amount of data collected globally, featured in the IPCC report and also in more recent research, indicates there is no real doubt for Western scientists about what is happening. Recently, it has also become clearer to the scientific community that the ability of the planet to keep cool and stable and keep greenhouse gases at bay is decreasing. There are several interrelated reasons for this, including the acidification of the oceans, thawing of permafrost, melting of ice sheets and glaciers, and deforestation. From the perspective of Chinese medicine, the planet is clearly experiencing excess heat on a global scale. As the research and data indicates, this increase in heat is occurring at the same time as the planet's ability to maintain coolness and climate stability diminishes. By overlaying a Chinese medical perspective onto this global issue, we can see that Western climate data indicates the planet is suffering from Yin-deficient heat.

Though the discussion surrounding the climate's dramatic changes might seem alarming coming from the often conservative Western scientific community, it has been twenty-five years since the first well-publicized, outspoken calls of public warning. With its publication in 1989, Bill McKibben's *The End of Nature* presented what was, at that time, the best available Western scientific understanding about what was happening to the climate, as well as different possible future climate scenarios. While it could be argued that there was not scientific consensus on climate change back then, there was certainly more than ample evidence to indicate serious issues were present that could create dramatic global consequences.

The year prior, the director of NASA's Goddard Institute for Space Studies and preeminent scientist James Hansen testified before the U.S. Congress that human activity was affecting the climate globally. As far back as 1988 he was saying that "I declared, with ninety-nine percent confidence, that it was time to stop waffling: Earth was being affected by human-made greenhouse gases, and the planet had entered a period of long-term warming." According to Hansen, with this warming comes an increase in global instability—heavier rains, more extreme flooding, and more intense storms.[12]

The situation of our global climate did not just appear magically overnight. As climate research indicates, it also does not resemble natural cyclical climate variations. Instead, it is the result of our individual and collective actions and inactions. It's also the result of the actions and inactions of our larger institutions, most notably in the United States and the Western industrialized world in general. Just like Jenny, we as a country and a culture have been too busy for too long, and many of the systems we have created have dramatically magnified this imbalance.

Once we recognize that climate change is the result of this imbalance, it is of vital importance to look below the surface of the condition, as a deeper diagnosis provides us the opportunity to treat where the symptom is coming from. By understanding both the multiple branches of the condition—the rising temperatures, decreasing forests, saturated oceans, thawing permafrost, and melting ice sheets and glaciers—and their underlying root causes, it's possible we can self-administer the medicine we need to address this sickness. Given the severity of this situation, now is the time to understand and treat the root causes of the crisis. By translating Western climate science to Chinese medicine, the severity of the situation indicates an urgent need to understand climate change as a symptom of bigger and deeper issues, within us, our country, and our culture.

CHAPTER 2

The Science of Climate Change

A Chinese Medicine Perspective

Both in my personal observations and the understanding of Western science, the climate is undergoing rapid and significant changes. When I first began writing this chapter in the spring of 2011, five months prior to the rain and flooding brought by Tropical Storm Irene, Lake Champlain had experienced unprecedented rises in water levels. In our northwest corner of Vermont, close to the lake, we had experienced unprecedented flooding.

It was a time that has made a lasting impression on me and many other Vermonters. With two months of torrential, sometimes non-stop rain, the lake rose to historic levels. The rain coupled with spring melt from a heavy winter snowfall led many of the rivers that feed into the lake to rise beyond their banks, and the lake itself was flooding. According to the Lake Champlain Basin Program, which monitors the lake and its water levels, the lake level exceeded the highest level ever recorded in 184 years of monitoring.[1]

The same forces of climate change that led to the flooding outside our clinic, described in chapter 1, had also been at play earlier that year. Vermont is known to experience some extreme weather, but local climate scientists described these as hundred-year floods—and we endured two of them in just a five-month period.

I remember several memorable sights from the spring of that year. One occurred in April when I was driving with my wife, Liz, in upstate New York to catch the ferry to Vermont. The road leading to the dock, where cars and people usually board the ferry, had water rushing over it from a nearby flooding stream. The stream usually ran parallel to the road for several miles, but in April, the stream had engulfed the road. This was a full breach of the riverbanks—the river was flowing rapidly along a path that just a few hours before was the road.

After finding a new route to the ferry, we arrived at the dock and saw a sight that would become a familiar one for the rest of that spring: the ticket booth was submerged halfway under water. The loading area was under about four feet of water. Once we found our way across the lake, on the Vermont side there was a similar scene: several feet of standing water covering the ferry dock.

The water from these spring floods would stay for several months, keeping the dock and ticket houses on both sides of the lake—as well as large stretches of Vermont and northern New York—submerged under several feet of water until it gradually receded by early summer. And while we may wish that this flooding was an isolated event and hope that what Liz and I saw was anecdotal, in reality it was part of a clear pattern of rapid, wide-reaching climate change. As we'll discuss later in this chapter and throughout the remainder of the book, the dramatic storms and severe flooding we've experienced here in Vermont are part of a clear, global pattern. Using the insight of Chinese medicine, we can see clearly that what is happening in New England is connected to similar severe weather around the planet.

If we look back over the past twenty-five years, in hindsight it makes sense that the planet is warming due to the gases we have been emitting. The basic science of global warming, equivalent to the introductory basics of Yin and Yang, explains that the gases we emit from driving, flying, manufacturing, and agriculture stay in the atmosphere, allowing sunlight to pass through but retaining the energy reflected from the earth's surface. This occurs because there is a change in the oscillation of the light as it transitions from sunlight to reflected light. As summarized in Fred Pierce's *With Speed and Violence,* from the mid-nineteenth century lab work was performed to demonstrate that some gases, including carbon dioxide, were transparent to ultraviolet radiation from the sun but trapped infrared heat that radiates from the earth's surface. Pearce states, "Later dubbed 'greenhouse gases' because they seemed to work like the glass in a greenhouse, these gases acted as a kind of atmospheric thermostat."[2]

As early as 1894, the Swedish chemist Svante Arrhenius began to study and calculate the correlation between gases in the atmosphere and the effect on climate. Even today, with the use of several decades of supercomputer modeling, the work of this lone scientist is still a relevant prediction of how greenhouse gases affect climate.[3] While the past two decades have seen an enormous amount of research on climate change, the basic scientific underpinnings and generally accurate predictions date back to the end of the nineteenth century. Arrhenius's work has been so fundamental to our current understanding that he has been called the "father of climate change."[4]

Moving forward to more recent contributions, two particularly striking elements stand out when considering Bill McKibben's seminal 1989 work. The first is how, within the last ten to fifteen years, Western scientific consensus changed, from viewing global climate change as a distant possibility to a here-and-now reality. Second, severe climate scenarios that were considered in the late 1980s to be theoretically possible yet unlikely are now occurring.

As McKibben states in the preface of his most recent book, *Eaarth,* in which he addresses both elements, "The first point of the book is simple: global warming is no longer a philosophical threat, no longer a future threat, *no longer a threat at all.* It's our reality. We've changed the planet, changed it in large and fundamental ways."[5] In the first chapter, he provides a summary of some of the recent scientific and popular literature that details how all aspects of life on the planet are being affected—plants, animals (humans included), insects, land and water, and the global climate as a whole. He cites studies that indicate that from Antarctica to Greenland, from Siberia to the Amazon, and from North America to South America, whole ecosystems are being fundamentally affected.[6]

Before we start to discuss the details of the science of climate change, I want to acknowledge that there are lots of numbers and studies. All of these are necessary for us to have enough information to understand what's happening. As you'll soon see, there's a continuing stream of studies and research that shows that the climate is rapidly warming and destabilizing.

This mounting evidence can feel overwhelming, like a deluge of bad news when we look at it from our usual Western perspective. However, when we see the climate science through the lens of Chinese medicine, clear patterns begin to emerge. The holistic thinking of Chinese medicine allows us to recognize the connection between the data collected from the tundra, the tropics, and more temperate locales like the United States. And in Chinese medicine, if we have a clear understanding of where symptoms are coming from—whether it's in the treatment room or with the climate—there's the opportunity to administer the remedies to address the root causes. There's more than just the catastrophe of climate change; there's the opportunity as well.

Cutting Down the Yin of the Planet: Deforestation

Part of the important role forests and trees play is that they absorb the greenhouse gases we create. They absorb carbon monoxide and carbon dioxide and, in turn, provide the planet with oxygen. From the Chinese medical view, part of what trees and forests do is take in the heat we release—heat that is mostly the result of the way we have been living. Trees have the ability to absorb heat because, compared to us humans, trees are recognized to be Yin in nature. While humans can move around—indicating more Yang—trees remain rooted where they are—indicating more Yin.

To understand the nature of trees, think about your daily life and compare it to that of a tree. If you are like many of us in the United States, you move around a great deal throughout your day—traveling for work, to school, to get something to eat, to see family and friends. For all of those hours that many of us are moving around and getting things done, trees are standing still, rooted into the ground. For Chinese medicine, trees and forests are not only Yin in their ability to help cool the planet, they are also Yin in their stationary nature. Forests and trees are important remedies for the symptoms of climate change in that they can absorb greenhouse gases, and their stillness and rootedness can also teach us important lessons about our overly busy, unrooted lives.

With the Yin nature of trees, understanding the speed at which we have been cutting them down is particularly significant. The staggering rate of global deforestation has been well documented for many decades. The United Nations' Food and Agriculture Organization (FAO) has monitored the world's forests at five- and ten-year intervals since 1946. Using data from over 230 countries and areas, with information from over nine hundred contributors using a wide variety of variables to measure forest health, the FAO describes their most recent assessment as the most comprehensive to date.[7]

While the assessment states that the rate of deforestation does appear to be decreasing globally, it demonstrates that "around [32 million acres] of forest were converted to other uses or lost through natural causes each year in the last decade compared to [40 million acres] per year in the 1990s."[8] This translates to 32 million acres of forest lost each year over the last ten years, or 320 million acres total in the past decade. It also means there has been 40 million acres lost yearly over the 1990s, or about 400 million acres in that decade.

Simple math shows that about 720 million acres have been deforested in the past twenty years alone. Translating this massive number into something we can comprehend, 720 million acres equals just under 1.1 million square miles of forests lost—approximately the size of California and Texas combined, or the size of France and Sweden together.

According to the report, forest planting and natural forest growth have reduced the net global effect of deforestation,[9] but planting trees is not the same as replacing the complexity and resilience of long-established forests. In terms of global climate change, part of what trees and forests provide is their ability to hold carbon that would otherwise be released into the atmosphere and contribute to warming. When deforestation occurs, eventually the greenhouse gases the trees were retaining is released. This effect is not minor by any means—the IPCC estimates that deforestation is the third-largest global contributor to greenhouse gas emissions, behind only energy production and industry. Given that the FAO estimates deforestation contributes seventeen percent of the total gases emitted, just the release of greenhouse gases from cut forests, which doesn't include the amount of carbon they would have stored, surpasses the total effects of all transportation, residential, commercial, service industry, and waste emissions, including that from landfills.[10]

By looking at the global condition of forests through what is happening around us, it's possible to see firsthand what is occurring worldwide. Back when I lived in Montana, it was clear that deforestation was not some distant possibility that might occur in some far-off place. Simple observation of the mountains in the Gallatin Valley showed a smaller, local picture that mirrored what was happening in forests around the world.

In addition to the rampant deforestation for timber throughout the United States, forests have also been fundamentally affected in other ways. Whole mountainsides—which, in a place like Big Sky Country, can span dozens of miles—are experiencing significant, and in some cases nearly complete, deaths of mountain pine trees. This is because pine beetles are now able to survive through the winter as the weather in the Rocky Mountains warms. A higher winter survival rate means more beetles, and more beetles means more pine trees are affected as the beetles bore beneath the bark looking for a place to live.

In a nearby state, "milder winters since 1994 have reduced the winter death rate of beetle larvae in Wyoming from eighty percent per year to less than ten percent.... Meanwhile, hotter drier summers have made trees weaker and less able to fight off the swarming beetles."[11] Providing context for how this affects the region's forests, environmental writer Jim Robbins summarizes that, "All told, the Rocky Mountains in Canada and the United States have seen nearly 70,000 square miles of forest—an area the size of Washington State—dead" in 10 years.[12]

From my Montana days, I can remember one hike I took with my wife, Liz, that offered a visual example of what was occurring within the region. A few miles outside Bozeman is a Forest Service road used by many people year-round for hiking, biking, and skiing. On a warm, clear summer day we set out from the trailhead parking lot. For the first few miles, the trail cuts into the mountainside and is bordered on the left by large ponderosa pines that grow over an understory of chokecherry bushes and wild forsythia. To the right of the path are openings that offer views of the nearby peaks as the trail follows the Sourdough Creek below. After a few miles, as we rounded a bend, we reached a clearing where we could see about a mile and a half into the distance. The view was a common one in the region—the pines were dying and their once-evergreen needles had turned brown. From what I could see that day, about three-quarters of the trees on the mountainside had been affected. Later hikes we took over the years in the Gallatin Valley told similar stories of large-scale and widespread die-off.

The importance of the death of local trees and expansive deforestation is highlighted by a recent comprehensive examination of worldwide forests. The data suggests that deforestation plays a much greater role in global warming than was previously recognized. The benchmark study, published in *Science*, provides the most accurate measure so far of the effects of deforestation on global warming and the amount of greenhouse gases absorbed by forests globally.[13] An international team of climate scientists combined data from 1990 through 2007 to create a profile of the role global forests have played as regulators of the atmosphere. The ongoing destruction of forests—mainly in the tropics for food, fuel, and development—emits 2.9 billion tons of carbon annually. This massive release of greenhouse gases accounts for more than a quarter of all emissions that stem from human activity. This updated number is over an eight percent increase from the FAO number cited above, a huge change when factored over a global scale for all greenhouse gas emissions.

The study also concluded that in addition to increased warming effects from deforestation, forests are responsible for more cooling effects than previously recognized when they are kept intact. Study coauthor Josep Canadell, a scientist at Australia's national climate research center, summarizes that

> this is the first complete and global evidence of the over-whelming role of forests in removing anthropogenic carbon dioxide…. If you were to stop deforestation tomorrow, the world's established and regrowing forests would remove half of fossil fuel emissions," and later describes these find-ings as "incredible" and "unexpected.[14]

Despite the extraordinary rate of deforestation, forests that are still intact today act as a tremendous depository for our emissions. New figures in the *Science* study reveal that the world's forests absorb 1.1 billion tons of carbon each year, the equivalent of thirteen percent of all the coal, oil, and gas burned in the world annually. Summarizing its significance for climate policy, Canadell says that the new data means that "forests are even more at the forefront as a strategy to protect our climate."[15]

An Embodiment of Coolant Is Warming:
Ocean Acidification

As the amount of greenhouse emissions is rapidly increasing and forests are being cut and dying, other factors that have historically kept the climate cool and stable are also changing. The oceans, an embodiment of Chinese medicine's understanding of Yin, may have reached their saturation point in their ability to contain greenhouse gases.

As noted in a variety of publications, including Yale University's Yale Environment 360 and those of business news company Bloomberg, another recent study in *Science* indicates that the world's oceans are acidifying faster now than they have over the last 300 million years.[16] In addition to affecting their capacity for holding greenhouse gases, this could also have serious consequences for many marine species and whole ecosystems. Summarizing the study, Bloomberg reports that in a review of hundreds of studies, a team of international scientists found that a steep rise in atmospheric levels of carbon dioxide has driven up oceans acidity levels ten times faster than the closest historical comparison—a period of acidification fifty-six million years ago that triggered a massive ocean die-off. The Bloomberg report continues that the oceans are vulnerable because they absorb excess carbon dioxide from the atmosphere, making the water more acidic.[17]

Acidification of the oceans, which results in a decreased ability to hold greenhouse gases, is particularly relevant as we now understand that the oceans have been holding a disproportionate amount of what we have been releasing. According to Hansen, "Earth's energy imbalance is deposited almost entirely into the ocean, where it contributes to iceberg and ice shelf melting.... It turns out that the lion's share of the excess incoming energy, about ninety percent, goes into the ocean."[18]

In Chinese medicine, water is a commonly used word to describe Yin, and the image of expansive amounts of water, such as the sea, is frequently referred to when describing Yin's cooling potential.[19] Part of what water does, literally and metaphorically, is cool things down. As the oceans are vast accumulations of water, and therefore vast accumulations of Yin, in Chinese medicine they are well suited to absorbing and holding emissions and heat.

As was the case with Jenny—whose Yin had been compromised, leading to hot flashes and other symptoms of heat—the ocean's ability to provide this cooling function has its limits. At some point, the ocean's ability, just like our own personal ability, to keep things cool and stable will eventually be compromised. The Chinese medical conception of Yin's limit for containing heat is no longer theoretical—we are now seeing this manifest within the oceans globally.

In addition to the ocean's decreased ability to absorb our emissions, a recent study in *Nature: Climate Change* indicates that the cooling effects of certain ocean organisms are also being compromised. Ocean acidification leads marine phytoplankton, which are microbes that live in sunlit water, to emit less of the sulfur compounds that contribute to helping keep the planet cool. Atmospheric sulfur is mostly derived from the sea, and phytoplankton produce a compound called dimethyl sulfide (DMS). Some of the DMS enters the atmosphere and reacts to make sulfuric acid, which then clumps into microscopic airborne particles. The particles help to seed the formation of clouds, which in turn help cool the atmosphere by reflecting sunlight that would otherwise reach the planet's surface and contribute to warming. The study estimates that this decrease in the creation of DMS alone could increase the global temperature by about 0.5 to 1 degrees Fahrenheit, leading to an overall increase in carbon dioxide levels.[20]

This might seem like an insignificant rise in temperature, but this recently discovered effect is factored over the whole planet. As with several other effects on the climate that we'll discuss later, its impact has not been factored into the projections by the IPCC and others. The combination of these effects are of tremendous importance. There is growing concern that we may be approaching a climate tipping point, where temperatures will begin to increase dramatically and rapidly.

The IPCC and others have suggested a goal of limiting the global temperature increase to about 3.6 degrees Fahrenheit. Seeing as how we have already increased the global temperature by slightly less than 1.8 degrees Fahrenheit, ocean acidification *alone* has the potential to raise the temperate by another 0.5 degrees, leaving us only 1.3 degrees from the target limit.

What Was Frozen Is Melting, Part 1: Permafrost

From the perspective of Chinese medicine, a frozen state is more Yin, while something that is thawed is more Yang. When something goes from being frozen to unfrozen, its temperature increases, indicating movement from the cool of Yin to the warmth of Yang. Similarly, Yin and Yang are also intricately interrelated: as Yang increases, Yin decreases, and vice versa. In other words, as something thaws, this indicates both an increase in Yang as well as a corresponding decrease in Yin. This dynamic can originate either from an increase in warmth—where Yang rises and Yin falls as a result—or from a decrease in cold—where Yin decreases, causing the Yang to increase.

This dynamic of warmth–Yang increasing as coolant–Yin decreasing is currently occurring in the oceans as well as on land in the Arctic. Looking again to Western science, ocean acidification and decreased DMS from phytoplankton is happening as permafrost melts in the region, both on land and in the oceans. Until recently, these permanently frozen parts of the Arctic have kept methane, a greenhouse gas, from being released into the atmosphere. Created as a result of anaerobic decomposition, methane has been stored in the ground in bogs and in the ocean floors, secured by a barrier of permafrost. Due to the warming global temperatures, however, permafrost is now melting, releasing methane that is potentially one hundred times more potent as a greenhouse gas than the more commonly discussed carbon dioxide.[21]

While there has been evidence of methane release since about 2008, as is happening globally with climate change, the process appears to speeding up.[22] Igor Semiletov, of the International Arctic Research Center at the University of Alaska, who has overseen climate research in the region and been studying the seabed for twenty years, has said that he has never before witnessed the current scale and force of the methane being released from beneath the Arctic seabed. Until relatively recently, it had been widely believed that any methane released in the oceans would mostly or completely dissolve before reaching the surface, limiting its warming effects—Semiletov's recent observations from relatively shallow waters reveal otherwise. In an incredible statement about the rapid changes, he told the UK's *Independent* newspaper that the plumes had increased from tens of yards in diameter to more than half a mile wide. He added that in a very small area, the researchers counted more than one hundred columns releasing methane directly into the atmosphere from the seabed. As a result, the concentration of methane was a hundred times higher than normal.[23] Just one year prior, Semiletov estimated that approximately eight million tons of potent methane was released from the region per year, but with the recently observed size and amount of the methane plumes, it's likely the actual number is significantly higher.

Similar to Semiletov's observations at shallower depths, Eric Kort and his colleagues from NASA's Jet Propulsion Laboratory have recently recorded elevated methane levels over the deep Arctic Ocean. Kort said,

> When we flew over completely solid sea ice, we didn't see anything in terms of methane. But when we flew over areas where the sea ice had melted, or where there were cracks in the ice, we saw the methane levels increase.... We were surprised to see these enhanced methane levels at these high latitudes. Our observations really point to the ocean surface as the source, which was not what we had expected. Other scientists had seen high concentrations of methane in the sea surface, but nobody had expected to see it being released into the atmosphere in this way.[24]

In looking to give it a monetary number, a recent study published in *Nature* investigated the possible economic costs of methane release from just one specific part of the Arctic Ocean—the East Siberian Sea off the northern coast of Russia. The study makes the extraordinary estimate that the total cost of the release "comes with an average global price tag of $60 trillion in the absence of mitigation action—a figure comparable to the size of the world economy in 2012 (about $70 trillion.)" This number includes a wide-range of global economic costs, including those from poorer health and lower agricultural production. The authors clarify that "the total cost of Arctic change will be much higher." In other words, the potential economic cost of the release of one greenhouse gas from one section of one ocean could be nearly equal to the total amount of income created by the entire planet in one year.[25] If the conclusions of the study prove correct, the cost of dealing with this region's methane release will financially overwhelm the global economy. Equally important is that the methane release from the Arctic region as a whole would have economic costs severely higher than that of the East Siberian Sea alone.

Looking to the effects from the land, melting permafrost within bogs has similarly been found to be contributing to the warming process. There are reports and firsthand accounts from scientists that the bogs in the Arctic region are melting at unprecedented rates.[26] Paralleling what is occurring in the oceans, a study in Siberia indicates that a massive release of carbon is possible due to melting permafrost. Summarizing a recent study, Michael Marshall states in *New Scientist* that "we are on the cusp of a tipping point in the climate. If the global climate warms another few tenths of a degree, a large expanse of the Siberian permafrost will start to melt uncontrollably. The result: a significant amount of extra greenhouse gases released into the atmosphere."[27]

Another recent paper also indicates that there appears to be much more greenhouse gases than previously estimated in frozen Arctic soil, and a higher likelihood that they will be released. As published in *Nature*, the title and the description are revealing: "High Risk of Permafrost Thaw," "Northern soils will release huge amounts of carbon in a warmer world." The article begins, "Arctic temperatures are rising fast, and permafrost is thawing. Carbon released into the atmosphere from permafrost soils will accelerate climate change.... Our collective estimate is that carbon will be released more quickly than [other] models suggest, at a levels that are cause for serious concern." To put the amount of carbon held in the permafrost into perspective, the article states that

> the latest estimate is that [the Arctic] northern soils hold about 1,700 billion tonnes of organic carbon—the remains of plants and animals that have accumulated in the soil over thousands of years. *That is about four times more than all the carbon emitted by human activity in modern times and twice as much as is present in the atmosphere now* [emphasis added]. This soil carbon amount is more than three times higher than previous estimates.

In terms of how much of this may eventually end up in the atmosphere, the authors state that based on different warming possibilities, "The estimated carbon release from this degradation is thirty billion to sixty-three billion tonnes of carbon by 2040."[28]

To give these enormous numbers some context, the most recent Environmental Protection Agency report states that the *total* emissions for the *entire United States* was 6.7 billion metric tons in 2011.[29] Averaging the potential carbon released from the Arctic permafrost over thirty years, this equals about one to two billion tons of carbon per year. In other words, the release from the permafrost may be equivalent to fifteen to thirty percent of the entire yearly U.S. emissions.

What Was Frozen Is Melting, Part 2:
Ice Sheets and Glaciers

As the permafrost melts, which is a decrease in Yin, and methane is released, which has a warming Yang effects, a similar dynamic is occurring elsewhere in the region. Similar to the role permafrost plays in climate stability, ice sheets and glaciers also have a cooling and stabilizing effect. In contrast to exposed soil, which absorbs warmth from sunlight, ice sheets and glaciers reflect sunlight back into the atmosphere, helping to cool the planet. As ice sheets—slabs of ice that are partially in water and partially on land—and glaciers—ice that is wholly on land—continue to melt and break away, sunlight is now being absorbed by the newly exposed land and water, contributing to the warming process.

In *The Independent*, Steve Connor cites a study in *The Journal of Glaciology* to describe clear global trends in Canada, the Himalayas, Greenland, and Europe. Other recently published articles tell similar stories about Europe, the Andes, Antarctica, and the United States:

- In **Canada**, glaciers in the western part of the country, which store an estimated 1,625 square miles of ice, are melting quickly and may completely disappear by the end of the century.

- In the **Himalayas**, hundreds of meltwater lakes have appeared in recent years on the surface of glaciers, indicating sustained glacial melt. Over a period of forty-eight hours, one such lake discharged approximately 27.7 million gallons of water, equivalent to forty-two Olympic-size swimming pools.

- In **Greenland**, computer models revealed that the melting in 2011 was the third most extensive melt on record since 1979, when recording first began, lagging behind only 2010 and 2007. In 2011, more ice melted than was formed by fresh snowfall.

- In **Europe**, glaciers in the French Alps have lost a full twenty-five percent of their area in the past forty years. For example, the ice fields around Mont Blanc and the surrounding mountains covered approximately 235 square miles in the 1960s, but by the late 2000s the area had decreased to about 172 square miles, about a twenty-five-percent reduction.[30]

- Another recent study published by the European Geosciences Union indicates that the glaciers in the tropical **Andes** have been melting at an unprecedented rate, one that hasn't been seen in the past three hundred years. Glaciers in the mountain range have shrunk by an average of thirty to fifty percent since the 1970s, according to Antoine Rabatel, researcher at the Glaciology and Environmental Geophysics Laboratory in Grenoble, France.[31]

• In **Antarctica**, new research indicates that the already significant rates of melting may have been significantly underestimated. A recent article in *The Independent* indicates that this underestimation in Antarctica was due to faulty data collection.

> Temperatures in the western part of Antarctica are rising almost twice as fast as previously believed, adding to fears that continued thaws are causing sea levels to rise…. In a discovery that raises new concerns about the effects of climate change on the South Pole, the average annual temperature in the region has risen by [4.3 degrees Fahrenheit] since the 1950s, three times faster than the average around the world.[32]

This is almost twice as fast as was previously believed for the rate of warming for the region. These accelerated rates of rising temperatures will also accelerate the rate of glacier and ice sheet melting.

• In the **United States**, there are numerous similar reports. A recent piece in the *New York Times* on dramatic melts in Alaska describes the effect on the town of Juneau. In the summer of 2011, sudden floods bursting from glaciers began occurring. "In that first, and so far biggest, measured flood burst, an estimated ten billion gallons gushed out in three days, threatening homes and property…. There have been at least two smaller bursts" since then.[33] The U.S. Geological Survey of the Department of the Interior cites research that describes occurrences in other parts of the United States:

In Glacier National Park (GNP) in Montana some effects of global climate change are strikingly clear. Glacier recession is underway, and many glaciers have already disappeared.... It has been estimated that there were approximately 150 glaciers present in 1850, and most glaciers were still present in 1910 when the park was established. In 2010, we consider there to be only twenty-five glaciers larger than twenty-five acres remaining in GNP. A computer-based climate model predicts that some of the park's largest glaciers will vanish by 2030.[34]

A recent *National Geographic* article tells a similar story and includes a comparison of older and current pictures of GNP to demonstrate the rapid loss of the park's namesakes.[35]

While living in Montana, I also witnessed these changes that may someday lead Glacier National Park to have to change its name. In 2006, carrying backpacks filled with the food and bedding we would need for a four-day trip, my family and I hiked along the Highline Trail, venturing into some of the most stunning landscape I have ever seen. With a view of hundreds of miles of peaks and valleys that dramatically rise and fall, it's clear why this area is called Big Sky Country. Out in this wilderness, the sky feels enormous, with its clear, dry air and unobstructed views. In addition to the breathtaking beauty, not far from where we set out on the trail, we experienced another reminder of the area's wildness: we had an up-close and personal encounter with a family of mountain goats on the trail.

The mountain goats seemed to be using the four-foot-wide trail because the sharp drop-off below left no other route for them to safely travel across the mountainside. After a long moment of tension and stillness, when we were only a few feet apart and cautiously looking at each other, the family of goats darted off the downhill side of the trail. They crashed through the small, dense bushes and a moment later reappeared a few feet down the trail, and ran off.

Once we arrived at the chalet, it wasn't difficult to see already what the U.S. Geological Survey and the *National Geographic* authors would write about several years later: large areas of the surrounding peaks and valleys had changed. The increasing temperature in the park, which had risen about 2.3 degrees Fahrenheit since 1900, almost twice as fast as the global average,[36] meant there were less white glaciers and more gray rocks and green shrubs. Not only did the melting of the glaciers mean the loss of the iconic nature of the park, it also means more sunlight is absorbed by the soil rather than reflected back into the atmosphere, contributing to warming.

Hansen describes the issues we saw in the park—warming temperatures melting glaciers, resulting in more warming temperatures—as "amplifying feedbacks that were expected to occur only slowly [but] have begun to come into play in the past few years." Significantly, he also notes that "these feedbacks were not incorporated in most climate simulations, such as those of the IPCC. Yet these 'slow' feedbacks are already beginning to emerge in the real world."[37]

Seeing the Forest for the Trees:
A Chinese Medicine View of Climate Science

When taken only at face value, the science of climate change can seem overwhelming. It can feel as if there is too much data about too many issues to make sense of it all. We might feel that too many things are happening at once, and too many feedback loops are occurring simultaneously for us to be able to understand the process that we've started. But just as in the treatment room, where it's not only possible to treat multiple physical symptoms simultaneously but address mental and emotional issues as well, Chinese medicine can help us clarify what we are seeing and allow a clear pattern to emerge. As discussed earlier, the long history of Chinese medicine is based on inductive thinking, which allows us see patterns and tendencies in the world around us.

From the view of Western science, forests absorb greenhouse gases and help to cool the planet. When there is large-scale deforestation from trees being cut or dying, this cooling effect is decreased, which is a decrease of Yin. The greenhouse gases the trees once absorbed are also released during deforestation, which contributes to a warming effect, which is an increase of Yang.

As the oceans acidify, they lose their capacity to store our emissions. This is another loss of coolant, another decrease in Yin. Ocean acidification also leads phytoplankton to emit less of the sulfur compounds that help seed cloud formation, which is another loss of a cooling effect, another decrease in Yin. Fewer clouds allow more sunlight to reach the surface, which contributes to more warming, which is an additional increase in Yang.

The melting of frozen soil in bogs and the ocean floor can be understood as the movement from a colder, Yin state to a warmer, Yang state. In other words, this is another decrease of Yin and an increase of Yang. As the permafrost melts, the potent greenhouse gas methane is released, again contributing to warming and increasing Yang. As glaciers and ice sheets melt, they also move from a more Yin, frozen state to a more Yang, liquid state. The resulting exposed soil and water absorb sunlight that would have been reflected, contributing to warming and an increase in Yang.

Looking at climate research through the lens of Chinese medicine, a clear pattern emerges. Together, the loss of forests, the acidification of oceans, and the melting of permafrost, glaciers, and ice sheets tell us a story of decreasing coolant and increasing heat. Rather than being separate issues, the situation of trees, ice, and oceans together depict a clear picture of a planet that is warming as its cooling capacities are decreasing. From the view of Chinese medicine, the cumulative effect of these symptoms tells us that the planet has a clear diagnosis: Yin-deficient heat. This increasing warmth with decreasing coolant is the cycle in which heat increases, cooking off the coolant. This decrease of Yin creates an additional increase in heat, which then cooks off more Yin.

To arrive at a clear diagnosis of what is happening to the climate is crucial. As in the treatment room, where acupuncturists and herbalists are looking to treat the specific root causes of various conditions, having a clear understanding of the global climate allows us to choose the most appropriate remedies for the imbalances we've created. When treating headaches, for example, there is no one-size-fits-all approach in which the same herbs and the same acupuncture points are used to treat the discomfort. Similarly, it's much more likely that we will be able to effectively address climate change if we understand the underlying issues rather than just the more superficial symptoms.

In addition to providing a clear diagnosis of the root causes, Chinese medicine can also help us understand the relationship between what occurs in the global climate and what occurs within us as individuals. The long-established connection in Chinese medicine between the big picture and the little allows us to see the clear connection between the microcosm and macrocosm. In doing so, this inductive understanding can help us discern the appropriate set of remedies to treat the sickness of climate change. Chinese medicine can help us see how basic assumptions about our lives contribute to our individual and collective imbalances that are being mirrored in the condition of the climate. Understanding Yin and Yang provides us the opportunity to self-administer the medicine needed to treat the causes of increasing heat and decreasing coolant, which is creating a rapidly destabilizing planet.

CHAPTER 3

The Meaning of Climate Change

The Personal Level

Yin's fundamental importance to a stable climate is mirrored by its importance to our own health. Physically, mentally, and emotionally, our Yin provides us with coolant that prevents overheating. Knowing that what manifests in nature is a reflection of what manifests within us, one way to address climate change is to be vigilant about the balance of our Yin and Yang. As we'll discuss later in this chapter and throughout the rest of the book, the heating of the planet—an increase of Yang—and its decreasing ability to hold greenhouse gases—a decrease in Yin—is being mirrored in our own internal condition. Climate change is not just happening in the world around us; climate change is also happening within us.

In addition to providing coolant, Yin provides a buffer against pathology. When we have enough Yin, sicknesses of all types can be prevented from appearing. In Chinese medicine, Yin and the fluids of our body are where a wide variety of issues that could develop into sickness can be stored.[1] Not only can viruses and bacteria that we haven't been able to expel be held, unresolved mental and emotional issues can accumulate here as well. Put another way, the Yin and our fluids have an ability to keep sicknesses of all levels in latency.[2] When these decrease, or when a sickness becomes too strong, sicknesses can begin to appear. Within us individually this can be for years and decades, and with the climate it has likely been for generations and centuries.

On a global scale, the Yin of the planet has served as a buffer against the greenhouse gases we have been releasing, as well as buffer against the consequences of these emissions. As previously discussed, in Chinese medicine, water is synonymous with the cooling quality of Yin—the oceans, as huge accumulations of water, are an equally large accumulation of Yin. Both Western scientific data and the inductive thinking of Chinese medicine agree that the oceans have a cooling effect on the planet. From the Chinese medicine perspective, the size and depth of the oceans suggests that they can buffer us against the effects of a great deal of unsustainability. At some point, however, even these global accumulations of water and Yin reach their limits, and their ability to keep in latency and the consequences of enormous amounts of greenhouses gases will be lost.

Frozen water—permafrost, glaciers, and ice sheets—is in even more of a Yin state, in that it is colder and less mobile. From both Western and Eastern perspectives, it makes sense that together these different forms of water and different forms of Yin have a cooling and stabilizing effect on the climate. Similarly, trees are also more Yin, as they are rooted to the ground, unlike us more Yang humans who are constantly moving around. The ability of forests to hold greenhouse gases is a more Western, contemporary way of describing the more traditional, Chinese understanding of the Yin nature of trees. As can happen within us individually, the planet's Yin influences—oceans, permafrost, glaciers, ice sheets, and forests—have been compromised. As their cooling capacity wanes, the warming consequences of our actions are no longer stored away and hidden. A clear and literal example of this loss of latency is the methane plumes that now dramatically bubble from the ocean floor to the surface of the ocean. Something that was once hidden under permafrost is now being visibly released into the atmosphere.

With methane bubbling from the ocean floor, there is a great deal of discussion about the policy issues of climate change. These including changes we need to make to our economic, energy, and agricultural systems. Undoubtedly, reorienting our economy away from greed and toward sustainability, a rapid and complete introduction of sustainable energy, and consuming more local, organically grown food are of vital importance. But before we look at these larger-scale issues that can be addressed at the policy level, let's first look at a more immediate and personal level of change—that of our own internal condition. Rather than thinking of climate change as something happening *out there*, let's consider how climate change is happening *within us*.

Instead of seeing climate change as a catastrophe, let us instead consider climate change as an opportunity. Yes, the planet is warming quickly; in fact, much faster than previously estimated. Yes, forests are being cut down at an alarming rate. Yes, permafrost, glaciers, and ice sheets are melting. And yes, methane is bubbling up from the ocean floor. As we look at the data of climate change, let's also consider what these facts mean in terms of a Chinese medicine understanding of what's happening.

Values That Feed Global Warming

We have seen that the global climate's dynamic of rising temperatures coupled with decreasing cooling effects is strikingly similar to Chinese medicine's concept of Yin-deficient heat. In looking at this excess of warmth, a basic question is: *Where is it coming from?* The simple answer is greenhouse gas emissions. This raises another question: *Where are these emissions mostly coming from?* The answer is that on a per-person basis, they're coming from Western industrialized countries, and the United States in particular. Though China recently became the largest overall emitter due to its large population, the United States is a close second and continues to top the list in terms of greenhouse gases per

capita of large countries. While there is some variation, Western industrialized countries are the largest overall emitters of greenhouse gases.[3]

In order to fully understand what is creating these emissions,
we need to look at our basic assumptions about the world. To begin
this process, consider the following questions, which are intention-
ally open-ended and vague. Don't think about the question or your
answer; just see what pops into your mind.

Ask yourself: Is it better to do something or not do something?

Is it better to have more or less?

Is it better if something is new or old?

The right or best answers may seem self-evident, or perhaps we
think these are funny or nonsensical questions. However, by using Chinese medicine as a guide, our responses to simple, open-ended questions
can reveal a great deal about who we are individually and collectively,
how we live our lives, and what we value. And this self-examination can
help us understand how we arrived at our era of climate change.

So many of the basic beliefs that we in the United States take as
unquestioned truths—assumptions about life, success, and how things
should be—are based on the values expressed when we answer these
questions. Just as Yin and Yang show up in nature and in the climate,
they are fundamental parts of us as well. Just as on the scale of the
climate, heat and warming are Yang, and coolant and cooling are Yin,
these same forces appear in all aspects of our lives.

Many of us in the United States, if we were being honest, would
have answered that it's better to do something, have more, and have
something new. We might even think that the answers are universal
or that everyone would respond the same way. But rather than being
self-evident, our answers actually speak to our individual and cultural
condition. When we believe that *doing* is better than *not doing*, we are
orienting toward the Yang. When we say that *more* is better than *less*,
we are also orienting toward the Yang. When *new* is better than *old*, we
are again orienting toward the Yang.

Of course, it's not just cultural coincidence that so many of us believe these Yang parts of life are better than their Yin counterparts. We have been encouraged and even taught to believe these things to be true. As we'll discuss later, we have built whole economic and social systems based on these beliefs. To help perpetuate these systems, we've been conditioned to think a Yang-centered approach is the best way, or even the only way, to live life; doing over nondoing, more over less, new over old. The issue is not that the Yang aspects of life are an inherent problem; the issue is one of balance.

When considering how to address our rapidly warming planet, recognizing the value of Yin is paramount. There is a real and tangible connection between what we value and the condition of the climate. Just as the climate is warming, we individually and collectively often overvalue the Yang. And just as the climate's capacity to keep itself cool is compromised, we often undervalue the Yin.

It is also not a coincidence that Americans are a leading contributor to greenhouse gas emissions, as we often strongly favor doing over nondoing, prefer more to less, and favor new over old. The consequences of our individual and cultural overvaluation of Yang and undervaluation of Yin has reached such an epidemic level that they are now affecting the entire planet.

Of course, some of the consequences of our overemphasis on doing, more, and new are straightforward. When we are constantly driving or flying, we emit more greenhouse gases. When we desire more things or the latest gadgets, the manufacture and shipping of these goods contribute to global warming.

However, there is more to it than just the practical consequences of Yang excess. There are certain thoughts and beliefs that have helped make this excess possible. If we want to address climate change at its roots by listening to the message it's telling us, we must reorient ourselves toward the Yin on all levels, including in our own lives. We must recognize the value of not doing, of less, and of the old if we want to know *within us* what climate stability would look and feel like in the world *around us.*

How can we possibly hope to create a sustainable country if we do not have a balance within ourselves of something as fundamental as Yin and Yang? How can we hope to decrease greenhouse gas emissions, stop deforestation, and halt the melting of permafrost, ice sheets, and glaciers if we're not clear how these same dynamics play out in our lives and our view of the world? My simple answer is that we won't.

Without an adequate appreciation of what brings long-term balance and well-being, we are very likely to continue in the direction we're headed. Without a way to understand what promotes health in our own lives, our country, and our environment, we are likely to continue down the path that has brought us to our present circumstances. This cultural inertia is leading us down one path, and Chinese medicine can help us see the other choices that are available.

The Stories We Hear and Tell

To examine our Yang overemphasis from a different perspective, think about what we hear on TV and the radio, or read about in magazines and newspapers: we often hear about Yang. Media often tell us about people who make lots of money, who make loud and brash statements, and about the newest and fastest products. But why do media outlets and publishers assume that this is what we would like to hear about? Why are these stories newsworthy?

We might assume that new things, loud statements, and making lots of money are inherently more interesting. Many of today's headlines exclaim things like "Actress Makes $20 Million for Latest Movie" or "Company Releases Fastest Cell Phone." These are examples of our Yang emphasis on more money, more speed, and new things. What if we saw more headlines like "Local Family Has Wonderful Weekend without Phones and Computers"? Or "Woman Describes Sitting in Silence for a Month"? That we see and hear much more of the first type of headline and much less of the second is not a trivial issue to climate destabilization. It exemplifies our overemphasis on *more, faster, and doing,* which is our focus on Yang. Our orientation toward these stories and these values is an orientation toward the Yang, which is the very same orientation that is helping to create a warming planet.

In our era of a rapidly warming planet, it's important to understand that Yang remains fundamental to our health and the well-being of the people around us; the issue is always one of balance. For example, there are people in the United States and around the world we really do need more, which indicates the need for Yang. There are many hungry and homeless people in the United States who need more food and more shelter. There are also many families who need more money to pay rent, buy food and clothes, and pay for health care. There are even more people in other countries who need more money and goods for their survival. We also need new technology because advances in solar and wind power, electric vehicles, and public transportation can help to limit or eliminate greenhouse gas emissions. But these new things will not save us from our current condition because in our era of Yang excess, more Yang is not the long-lasting remedy. In our times of climate change, Yang certainly has its place. As with our own lives, however, the issue is one of balance.

The Importance of Balance

It is of fundamental importance to our health and the health of our climate to avoid extremes. In terms of Yin and Yang, this includes not forsaking one end of the spectrum for the other. It is not a sign of well-being to go from thinking that Yang is "good" to seeing Yang as "bad." Going from one extreme to another does not create balance. Health, both in ourselves and in our country, is the middle ground between rest and activity: having the pendulum swing from "all action all the time" to "all rest all the time" is not a sign of well-being.

This place of balance is not a fixed location but rather is continuously moving based on the many variables that create health. Personal health and environmental sustainability are not specific dots on a static two-dimensional map of well-being. Balance is part of a dynamic process that, by its nature, changes constantly. For example, in writing this book, it has taken a sustained effort of directed Yang as I have been actively researching and writing about the blending of climate change and Chinese medicine for several years. For a decade I have been talking and thinking about these Western and Eastern ideas and how they fit together. If it wasn't for Yang, this book would not exist and you would not be reading these words today. Yang is the activation of ideas and thoughts, and the ability to conceptualize, start, and finish a project such as writing a book.

But what creates the Yang? Healthy actions—and in fact all of life—comes from the Yin. Put another way, the Yin *creates* the Yang while the Yang *is an expression of* the Yin. The rest, relaxation, and contemplation that are the Yin allow us to promote and sustain health.[4] This is similar to the explanation by Western biology that humans evolved from creatures that originally lived in the oceans. From a more literal Western biological view, human life began in water, and from a more metaphorical Chinese medical perspective, healthy action originates from Yin.

When there is harmony, Yin balances the Yang and Yang balances the Yin. It may seem paradoxical to our more usual Western thinking that Yin and Yang work with each other. We might be more comfortable with the idea that doing and nondoing are opposites, or that there is a struggle between the old and new. What Chinese medicine can teach us is that Yin and Yang are not oppositional but rather complementary forces. Yin and Yang are in no more conflict than the light of day is fighting the dark of night or the warmth of summer is battling with the cold of winter.

The importance of Yin to a healthy life is also apparent in the deep-reaching philosophical ideas of Chinese medicine. In addition to being associated with rest and relaxation, Yin is also understood to represent wisdom. In our own lives, this includes understanding what promotes long-term health on all levels. Collectively, this involves a deep understanding of long-term environmental sustainability. Without an embodiment of the wisdom that Yin provides, it's not likely we'll live a healthy life individually or create long-lasting environmental sustainability culturally.

The Yin-Yang symbol is a visual representation of what Chinese medicine also articulates in words. The black of the Yin and the white of the Yang transform into each other. The colors are not divided neatly down the middle but instead flow into each other. Even in our era of climate instability, Yang is not bad. After all, Yang is sunshine, warmth, summer, activity, and growth. These are fundamental aspects of nature and essential parts of our lives.

Rather than being separate ideas, or oppositional parts of life, Yin and Yang are intertwined. As the Yin-Yang picture demonstrates, the light that is the Yang becomes the darkness that is the Yin, and the Yin eventually transforms back into the Yang. That the visual representation of these ideas is a circle is intentional—as the white expands and begins to reach its fullness, darkness begins to appear, and vice versa. Just as night become day, and then again becomes night, the Yin becomes the Yang, which then again becomes the Yin.

Within the predominantly black Yin area of the circle, there is also a dot of white Yang. Likewise, within the predominantly white Yang, there is also a dot of black Yin. When there is balance in own lives, we will have a sense of engagement even when resting and a sense of relaxation when active. On a deeper level, the dots of color express that with personal health and environmental sustainability, the wisdom of the Yin is embedded within the action of the Yang. Likewise, within us, with the rest and relaxation of the Yin is the Yang engagement with the world.

One way of making sense of the vast amount of data on climate change is through understanding that our individual and collective imbalances have become so pronounced that they are affecting the climate of the entire planet. If we were aware of how to live a balanced life, and were acting on this understanding, the climate could not be in the state it is today. Put another way, what is happening to the climate is a direct reflection of what is happening within us.

Consumerism: A Symptom and Cause of Yin-Deficient Heat

From my clinical experience, it's likely that Chinese medicine practitioners in the United States will continue to treat a great deal of Yin-deficient heat. This is because greenhouse gas emissions can be seen as both a literal and metaphorical example of hyperbusy people, a hyperbusy country, and a hyperbusy culture. Given the vast amounts of data on climate change, why do we continue to travel and consume things in a way that has such a significant global impact? Put more simply, why do we continue to do what we are doing? One basic reason is that we collectively have too much heat and not enough yin.[5]

We are continuously encouraged to want more—more things, more money, more status, and more in general. So much of the entertainment and advertising we see encourages a sense of dissatisfaction with who we are and what we have. The sale of many products is predicated on feeding and even creating our sense of want.

Our Yin is our sense of contentment and peace. Yin is our feeling of being satiated physically, mentally, and emotionally. Yin is also what gives us the ability to say "that's enough, I don't need anymore." Yin allows us to be more satisfied with the things we already have and with life in general. With a planet that is warming due to our excess busyness, it is not surprising that so many of us feel so dissatisfied despite buying more and moving around at ever-faster speeds. Basic Chinese medical theory makes it clear that doing more is not likely to create a greater sense of peace. In a Yin-deficient culture, increasing our level of activity and owning more things is not the antidote. What, then, is the medicine for the sickness of too many people wanting too many things and being too busy for too long? *Doing less and wanting less, and being content and satisfied more.*

When it is no longer connected to what we really need to live, being busy and wanting things creates heat within us. As happens on an exceptionally hot summer day, this extra warmth is uncomfortable and creates a feeling of agitation. Over time, this excess stimulation can contribute to more heat, further increasing our discomfort. It's not surprising that buying things we don't really need doesn't bring lasting satisfaction. In fact, it does the opposite—it increases our sense of stimulation, which can cook off more of the Yin that would have provided a sense of satiation. This consumerism decreases our sense of satisfaction and contentment, which in turn leads us to look for more stimulation and excitement.

Many manufacturers, stores, and advertisers are more than happy to offer us a short-lived reprieve from this internal dissatisfaction by selling us the next great thing. In the short term, we can perhaps distract ourselves from the inherent unpleasantness of Yin deficiency and heat by buying superfluous things. Or perhaps we can create some short-term relief by distracting ourselves, either by overworking or entertaining ourselves. It's also possible we can reduce our discomfort for a short while by using various substances, both prescribed and unprescribed, that numb our senses.

But not surprisingly, once these distractions are gone, we are back to where we began—an overstimulated internal condition with a lack of internal peace. Our attempts at distracting ourselves are not only unsuccessful, they often contribute to the internal condition that created these desires in the first place. While wanting more is certainly not always pathological, and while there are people who really do need more, we have been encouraged to assume that if *some* of something is good, then *more* of it is better. Not surprisingly, Yin-Yang theory says otherwise. The inductive thinking of Chinese medicine allows us to see the manifestations of these misguided assumptions as part of the fuel that is feeding climate change. Not only does the Yang of more feed our desire for more material things, it can also compromise the satisfaction of having less that comes from the Yin.

Healing Climate Change Internally: Personally Decreasing Heat and Increasing Coolant

With the importance of appreciating that both the cause and the cure for global warming can start within, there are several ways we can balance our own Yin and Yang.

First, realize that doing is not better than not doing. There is a time to work and a time to rest, a time to be busy and a time to slow down. This is especially important for those of us already engaged in the work of addressing climate change. We must practice balance in our own lives if we are committed to climate sustainability. Being frantically busy, even in the quest to address climate change, is internally recreating the same dynamic that is rapidly warming the climate.

Second, appreciate that more is not better than less. Part of the consequences of this overemphasis on more is the increase in greenhouse gases from the manufacture and shipping of things we don't really need. However, always wanting more also diminishes our own sense of Yin contentment and increases our Yang desires. Continuously wanting more, even in the name of climate sustainability, is not likely to create a balance within us. How do we hope to create a cooler planet if we are overstimulated internally?

Third, realize that new is not better than old. We mistakenly assume at times that something new is inherently better than something long established. Whether it's a phone, technology, or a medical treatment, something that is new is not necessarily better than something that is old. Part of the antidote for the deeper causes of climate change is revaluing the traditional in all parts of our lives, and all parts of our country and culture. This includes the value of the millennia-old thinking of Chinese medicine, and acupuncture treatments and herbal prescriptions. In our era of Yang excess, we need to be particularly discerning about which new things we value and incorporate. One way to gauge the relevance of new things is to see if they have a Yin result. For example, new solar and wind technology—which is Yang—can lessen greenhouse gas emissions—which is Yin.

Fourth, cultivate an internal balance of Yin and Yang. Visit places that Western science says store greenhouse gases and that Chinese medicine describes as cooling. Go to the oceans and see what you can learn from the rising and falling of the tides. Sit next to lakes, rivers, and streams and investigate the nature of water. Go to the forests and listen to what is happening and not happening. Sit with your back resting against a tree and just *be.* Let the forest teach you about being rooted and staying still. These can all help you experience more deeply the Yin.

Fifth, consider acupuncture and Chinese herbal medicine. One of the best and most effective ways that I know to find balance is with the help of the diagnostic and treatment principles of Chinese medicine. One way to help the planet is to help ourselves. As we move closer to balance in our own lives, we will be better able to understand how to move our country and culture toward health. Chinese medicine can effectively clear out heat and bring in coolant, two things of vital importance in our era of overstimulation.

Sixth, start or maintain an internal practice. Chinese medicine has well-developed techniques to help us cultivate well-being. Practices like mediation, Tai Chi, and Qi Gong are all Yin by nature, as they involve sitting quietly and moving slowly. They can enhance our own health and help us cool down physically, mentally, and emotionally.

Seventh, to paraphrase Mahatma Gandhi, become the changes you wish to see in the world. If you hope to see a cooling planet, *become* more Yin. If you hope to see a decrease in greenhouse gas emissions, clear out excess heat and overstimulation *within yourself.*

Not only can the above suggestions help us experience internally what a balanced country and culture would look like externally, they can also help us find a deeper sense of well-being. In this time of potentially cataclysmic climate change, it is our health—specifically our own internal balance of Yin and Yang—that will enable the deep changes necessary for sustainability to occur.

In addition to the beliefs that contribute to a warming planet, what we eat and drink can also add fuel to the fire. If we want to treat where climate change is coming from, it's important to understand how our diet can contribute to an overheated condition, both internally and globally.

As we'll discuss next, something that's as ubiquitous as coffee can have very significant effects on our well-being and, in turn, on the health of the planet.

CHAPTER 4

Feeding the Fire of Climate Change

In order to understand what the symptom of climate change is trying to tell us, it's essential we understand that what's happening to the planet is happening to us. The macrocosm is a reflection of the microcosm—our warming planet is a mirror of our own individual and collective condition. Just as our assumptions about the world and our lives help warm the planet, what we ingest can also feed the fires of climate change.

In Chinese nutrition, substances we eat and drink are categorized based on their qualities. Rather than talking about things like vitamins and minerals with Western nutrition, Chinese medicine uses terms coming from nature to describe food. For example, some food is considered hot, some cold, some drying, some moistening. Some food brings energy up-and-out, some brings energy down-and-in. Mirroring and feeding our philosophical overemphasis on Yang is our regular ingestion of a hot, very stimulating substance—coffee.

Just because something is commonplace does not mean that it's healthy for us or the planet. Especially with our country and culture so out of balance that it's rapidly overheating the climate, many things that seem "normal" are in fact pathological. The amount of coffee we drink is one example. Working with many people who drink coffee reveals a clear pattern in its effect on our Qi. Coffee overstimulates the whole body, and its heat is of a nature that can be difficult to clear.

Many people who drink coffee are looking for a pick-me-up, especially in the morning. One basic reason for this is that so many of us are, at a very simple and very basic level, tired. Being tired in our overly stimulated, too-busy world shouldn't be surprising as working too much, being entertained too much, buying too much, and traveling too much wears us out. Rather than running on the fuel of Qi, many of us are running on the fumes of heat.

While drinking coffee might seem like a quick and easy fix, it doesn't really provide us what we are looking for. Just as shopping, being busy, and constantly moving around don't provide us with any lasting satisfaction, hot stimulants don't give us lasting strength. When we're tired, part of the remedy can be to increase the amount of Qi we have. Coffee does not provide us with this energy; instead, it gives us overstimulation.

The Nature of Coffee

Part of being amped up on stimulants like coffee is that too much energy is going up and not enough energy is going down. In terms of Yin and Yang, this is too much Yang—too much up—and not enough Yin—not enough down. As discussed in chapter 3, this is the very same dynamic as climate change, where the planet's heat and Yang are increasing as its coolant and Yin are decreasing. While we may feel more awake as a result of coffee bringing more blood and Qi to the brain, there is an unrooted quality to its effects. The way that coffee works is by strongly encouraging our Qi to rise upward. With the understanding that things are interconnected, we see that if we ask our energy to ascend abruptly, the balance of our descending energy can become compromised, leaving us overstimulated and ungrounded.

Another way to understand the difference between having an abundance of Qi and being overstimulated is to consider the quality of the energy. Abundant Qi produces the sensation of strength and clarity and simultaneously a feeling of peace and relaxation. When we drink coffee, we may feel strong and alert, but we are also likely to experience other things as well. We might feel agitated, restless, anxious, and unable to sit still and relax. If we drink coffee frequently, we might even feel these things with such regularity that it seems par for the course. We might assume that the effects of being overstimulated are an acceptable or even an inevitable part of having enough physical strength and mental clarity to get through the day. But just as there is nothing inevitable about a warming planet, there is nothing inevitable about being amped up.

Not much is written about coffee in Chinese medicine. This lack of information is particularly interesting considering that thousands of substances have been discussed in great nuance and detail in herbal medicine and nutrition over several thousand years. This includes substances from all over the world, from common plants like dandelion and burdock to more unusual things like deer antler and gecko. Historically, it seems that coffee was not considered useful as a food or medicine and was simply not worth writing about.

Well-known contemporary Chinese medicine practitioner and author Bob Flaws does write about coffee in his Chinese nutrition book *The Tao of Healthy Eating*. Author, editor, and translator of over sixty books on Chinese medicine, Flaws describes coffee as warming and upthrusting, converting our deep reserves of energy into short-term stimulation. As a diuretic and laxative, it can also drain Qi out of the body through an excess of elimination. Its overwarming quality also contributes to heat in the Stomach and Intestines.[1] In my clinical experience, the overheating quality of coffee often spread from the digestive system to other organs, contributing to inflammation throughout the body.

In addition to coffee's heat, my clinical experience demonstrates that it is also damp. To understand Chinese medicine's concept of dampness, think about phlegm. It's something sticky and often heavy that clogs things up and slows things down.[2] While heat is an excess of Yang, dampness is an excess of Yin. When you have a stuffy nose, for example, dampness and phlegm are obstructing your breathing. The oil on top of a cup of coffee is an indication of this dampness, and it doesn't appear just in the throat or Lungs, but can be anywhere in the body.

While the heat of coffee is stimulating and pushes things upward, the dampness slows and pushes things down. Together, these two effects create what Chinese medicine calls damp heat. To understand what Chinese medicine means by this term, think of a hot, humid summer day. The temperature is high and the air is thick and heavy. The heat creates stimulation, but the humidity weighs us down and makes us feel lethargic.

Dampness is the accumulation of too much moisture, a condition in which the potentially cooling and cleansing effects of our internal fluids have instead become imbalanced. As it's an accumulation of fluids, and as fluids are Yin, dampness is a Yin condition. Yin conditions tend to sink downward, just as Yang conditions such as heat tend to rise upward. Coffee's dampness—its Yin—can trap the heat—the Yang—that coffee puts into the body, encouraging the Yang to linger.

For anyone hoping that the stimulation of the heat is offset by its dampness, there's bad news: two pathological influences don't create balance. Instead, they create imbalances that can become layered on top of each other, making each more difficult to resolve. Dampness as an excess of Yin and heat as an excess of Yang can create real turmoil, in the digestive system in particular. The Stomach and Intestines can have a hard time trying to sort out and get rid of this simultaneous excess of Yin and Yang.

A helpful and instructive image of dampness trapping heat is that of a wet blanket on a bonfire. You can put the blanket—the dampness—on top of the flames—the fire—and the flame may go down, but the heat from the embers can linger. As it is heavy like a wet blanket, the dampness from coffee will often push the coffee's heat into deeper and lower parts of the body, particularly the abdomen.

Retaining both dampness and heat internally is like being outside on a hot and humid day all the time. Just as the humidity can make the high temperature feel much worse, dampness can magnify the effects of heat within us.

To continue exploring the effects of damp heat, let's look at another understanding of heat in Chinese medicine. Chinese medicine cites summer heat as the likely culprit in cases of heat stroke.[3] As the term suggests, summer heat stroke happens when we are exposed to too much summer sun and heat for too long. One of the contributing factors is high humidity: the extra moisture in the air during a humid day coats the skin and prevents us from sweating out the heat. Similarly, internal dampness can trap heat internally, preventing it from being cleared out of the body.

In describing the history of coffee's use, Flaws reports that when coffee was first introduced into Europe,

> there were prohibition movements and laws based on the recognition that coffee is powerful and not a wholly benign drug.... Its use is not very wise. It is my belief that if coffee were to be introduced to the West today as a new discovery, governmental agencies, such as the FDA in the United States, would restrict its use as a controlled substance.... except as a medicinal and in cases where the use of speed is warranted knowing full well the risks it entails, I believe coffee has no place in the diet of those hoping to be healthy. It is one of the few foods that I unequivocally deny to my patients.[4]

Clearly, Flaws's picture of the effects of coffee isn't pretty, especially considering its ubiquitous use in the United States. In a more contemporary, Western view, coffee is often presented as an energizing drink that has lots of potential health benefits.[5] However, what Western medicine describes as a possible benefit comes with a long list of potential problems from its overstimulating nature. From what Flaws writes and from my experience in the treatment room, it's clear that coffee is one of the most pathological substances we regularly ingest. The imbalance it helps create not only affects us personally, but it also contributes to our collective overstimulation, which in turn fuels climate change. Part of what drives so many of us to keep going, far beyond what the planet can sustain, is overstimulation from coffee.

The Effects of Coffee: The Personal

Because everyone is a unique individual with a unique balance of Yin and Yang, we'll all have a somewhat different response to coffee. My clinical experience, however, is that coffee clearly introduces a large amount of heat, which we in the West call inflammation when it manifests itself physically and overstimulation when it manifests itself mentally and emotionally.

For many of us, drinking coffee can produce a long list of symptoms. My experience in the treatment room suggests that the hot, upward-rising nature of coffee can weaken our descending energy. This excess Yang and lack of Yin energy can contribute to, or create, a wide range of symptoms, including:

- anxiety
- racing thoughts
- insomnia
- disturbing dreams
- headaches (especially migraines)
- acid reflux
- irritable bowel syndrome
- a wide range of stomach and intestinal symptoms
- night sweats and hot flashes
- fibroids of all kinds, including uterine fibroids
- growths of all kinds, including tumors

- a wide range of skin conditions including eczema

- arthritis

- a wide range of pain conditions, including fibromyalgia heart palpitations

- excess anger and aggression

- dizziness and vertigo

- lower-back and leg weakness and pain

- a lack of rooted energy in general

In fact, all chronic and acute conditions that involve heat from the Chinese perspective or inflammation from the Western perspective are likely exacerbated by coffee. The good news about this long list of symptoms is that I have seen all of them improve or go away completely with the elimination of coffee as part of Chinese medicine treatments.

One example of the lasting effects of coffee was the experience of a woman I'll call Marsha. She came into our clinic with a wide variety of symptoms, including insomnia, anxiety, and fatigue. During our first appointment, she described how most nights she was only able to sleep a few hours. She talked about how, in the middle of the night, she would wake up with her mind racing and be unable to fall back asleep.

Not surprisingly, she woke in the morning tired. Marsha also described how here fatigue lasted throughout the day and how she often felt pressure in her chest that accompanied a sometimes severe sense of anxiety. This anxiousness had gotten to the point where it was significantly affecting her work and her interactions with her family. The anxiety could come on quickly and, at its most severe, she had to be by herself and avoid all contact with other people.

As is common in our clinic, we talked about how the food we eat and the beverages we drink can contribute to our health and our sickness. And, as is often the case, we talked about how the stimulating nature of coffee was contributing to her internal condition. After several months of acupuncture treatments, taking herbs daily, and diet and lifestyle change, Marsha was feeling much better. She was sleeping soundly through the night and her energy was much improved. Her emotions were also much better, and the infrequent anxiety she did experience was much less intense and passed in a few minutes.

Marsha was coming in for treatments about once a month when I took her pulse and felt a dramatic increase in heat. I asked her if she had increased her exercise or had more stress in her life, as both can increase heat. After she said no to both, I asked if she had started drinking coffee, and again she said no. I continued to feel her pulses and asked her again about coffee, and she paused and with a somewhat startled look said, "I only had one cup of coffee about a month ago."

The fact that I could very clearly feel in Marsha's pulse the heat from a single cup of coffee a month after she drank it speaks not only to its very stimulating nature, but also to how it lingers in the body. As is common in many people who drink coffee, the heat and dampness were accumulating in Marsha's Intestines. As is part of its job, the Large Intestine in particular was trying to clear things out, but the dampness was affecting elimination. As a result, both heat and dampness were lingering there.

The Context of Coffee

To provide some context, it's not that a substance that is hot is necessarily bad. Just as Yang isn't inherently bad even in our era of climate change, a drink that lifts our energy upward is also not inherently pathological. Chinese herbalism has a long history of using warming herbs that enliven the Qi. For example, ginger and cinnamon bark are both considered hot, have a long historical use, and continue to be used regularly today in clinical practice.[6] As with the larger understanding of Yin and Yang, the issue is one of context.

In my clinical experience, addressing both the Yin and Yang together is usually the most effective approach, especially in our overheated country. As we talked about earlier, Yin and Yang are best addressed together because in the view of Chinese medicine, the Yin is the basis of the Yang. In other words, without enough Yin, you can't have adequate Yang. A common analogy for Yin's relationship to Yang is an oil lamp. The oil itself is the Yin, and the flame is the Yang. Without enough oil, the flame can't burn brightly. And when the Yin of the oil is gone, the Yang of the flame goes out.[7]

The issue with the coffee we drink is the context. The first part is that we drink coffee on its own, without a balance of other more Yin, cooling, and descending herbs. The second is that we live in a time when our lives are so amped up from producing things, buying things, and moving around that we are rapidly warming the whole planet. In our era of climate change, part of the medicine we and our planet need is less heat, not more. Regularly ingesting a hot, stimulating substance while living in an overly busy and overstimulated culture adds fuel to the fire. While eliminating coffee from our diets on its own isn't likely to address climate change, it's certainly an important step in the right direction. The third part is the amount of coffee we drink. As we'll see a little later in this chapter, we in the United States drink an enormous amount of coffee—billions of cups each year.

Talking in the Treatment Room about Coffee

At this point, I've spoken to several hundred patients in the treatment room specifically about the effects of coffee. From their responses and questions, I can guess what some of you are thinking.

You might be saying: *But I heard that coffee was good for you.* Coffee is clearly stimulating and does increase blood flow to the brain and other areas. It can provide a lift of energy, but many short-term fixes come with long-term costs. In addition to this heat, coffee is also damp, which can trap the overstimulation and make it harder to clear from the body. The excess rising Yang energy of coffee can also compromise our descending Yin energy, which is the very same dynamic happening to the climate. Just as the planet is warming as its ability to keep things cool and stable decreases, our bodies become overheated and inflamed as their ability to cool themselves is impaired. For us individually as for our planet, coffee causes a host of heat-related symptoms.

Some of you might be thinking: *Well, if coffee is hot, then drinking iced coffee will help cool it down.* In talking about the characteristics of food, Chinese medicine is describing the food's inherent properties, which are different from its physical temperature. Putting ice in a cup of coffee does not transform its fundamental nature; what it does do is make it more difficult to digest. When a food or drink is below room temperature, the Stomach first needs to warm it before it can start digesting. The extra Qi that this warming requires means there's less available for digestion, making the food-to-energy process less efficient.[8] As the digestive system gets bogged down from the cold, dampness is generated. One common metaphor in Chinese nutrition is that the Stomach is a cooking pot that needs to be maintained at a warm temperature to work effectively. The cold of ice in coffee can decrease the warmth that the body needs to maintain effective digestion, potentially creating more dampness.[9]

You might be thinking: *I drink decaf, so that makes it okay.* As with putting ice in coffee, removing some of the caffeine does not change the coffee's basic nature. Regardless of the particular level of caffeine, coffee is still coffee, which is hot and damp. In many cases, the chemical processes that are used to remove the caffeine add toxicity. But even if a more natural, water-based process is used, from a Chinese medical view, the nature of coffee doesn't change with its level of a particular chemical.

Another common response is: *I don't feel good when I don't drink coffee.* A major reason we drink coffee is that we're tired, and removing coffee's very stimulating effects can make us more aware of what's going on. In our era of a rapidly warming and destabilizing climate, it's essential that we see the situation clearly, both for us and for our planet. Just as there is a long list of issues associated with drinking coffee, there can be a long list of symptoms associated with stopping. These include headaches, constipation, upset stomach, fatigue, lack of mental clarity, irritability, and cold sweats.

The good news is that acupuncture and herbal medicine can be very helpful mitigating or eliminating these symptoms. Not only can there be a relatively easy transition with the help of Chinese medicine, but people often quickly realize that they have more energy and clarity soon after stopping. The process of looking clearly at our lives involves not only considering what we are willing to do, but also considering what we are willing to do without. In addition to its relevance to our own well-being, addressing our use of coffee to keep going during our era of overbusyness has real, global importance.

Some people think: *I like the feeling of being "up" from coffee.* That "up" feeling is the coffee's heat. This overstimulation mirrors the heat of climate change. Part of what coffee provides is a way to avoid actually experiencing our levels of Qi. In our era of constant demands and distractions, it's not surprising that so many of us are drawn to the Yang of being up rather than the Yin of being down. But as the Yin-Yang circle shows us, Yin and Yang are interconnected. It's the down of Yin that creates as well as balances the up of Yang. Without the downward, Yin feeling of rest, sleep, and feeling tired, we can't sustainably create the up, Yang feeling of strength and power.

Some of you might be thinking: *I put a lot of milk in my cup, so I'm not drinking too much coffee.* From the viewpoint of Chinese nutrition, anything more than an occasional small amount of milk can create dampness. For many people, too much dairy creates phlegm in the nose and Lungs. Not only can this dampness be in the respiratory system, but it can occur throughout the body.[10] As coffee can also contribute to this phlegm, adding dairy to coffee can exacerbate its ill effects.

Maybe you're like some people who have come to our clinic and say, *I just like the richness of coffee.* Luckily, there are many other rich drinks that don't have the same heat- and damp-producing effects as coffee. As we'll discuss a little later, well-steeped dandelion tea can be dark and rich and can help clear out the heat and damp of coffee.

Maybe you think, *I like the hot drink in the morning and the ritual of making it.* Warm, cooked food and drinks help promote healthy digestion, and a morning ritual for starting the day can also promote health. However, having these include coffee is not particularly healthy.

Some of you might think, *I'm supporting socially responsible companies when I buy fair-trade coffee.* While buying fair-trade products can certainly be better for the people who grow it and for the planet, coffee remains a very stimulating drink. It's not very likely that we'll be able to address the root causes of climate change if we continue to be overstimulated internally. There are lots of other thoughtful, socially responsible drinks to choose.

Having worked with hundreds of people who have stopped drinking coffee, I realize that it is not always an easy process. If you want to stop or at least give it a try but are having a hard time, my recommendation is to find a good Chinese medicine practitioner. Rather than creating overstimulation, acupuncture and herbal medicine can provide what many of us are looking for when we drink coffee, namely Qi. A well-trained practitioner can also help clear out the heat and dampness that has likely accumulated and help replace it with physical strength and mental clarity.

The Effects of Coffee: The Collective and the Climate

Not only is coffee overheating us individually, it is a dietary contribution to our warming planet. Once again, what is happening within us is being mirrored in our culture and in the climate—the microcosm is a reflection of the macrocosm. Even if it's well-intentioned, being amped up still means that you're amped up. Drinking coffee to fuel your attendance at the next big climate rally is the same energetic condition as drinking coffee to fuel the pursuit of buying or selling the next big thing. Depending on your perspective, you may think that one is more important than the other. But the nature of coffee is the nature of coffee, which is hot and stimulating.

When you're overstimulated from heat, you can feel compelled to keep doing things. It can feel impossible to sit still or take a break because your internal condition is telling you, "Keep going!" You don't have to be a practitioner of Chinese medicine to recognize the effects of coffee. You can see the change in people as they drink it. They often get redder in the face, talk louder and faster, move around more, and become more assertive. Often their voices get louder and sometimes they start to shout. Many times, if someone has two or more cups relatively quickly, you can see these changes right away. While none of these responses necessarily indicates a problem—there are people who could use more of all of these things—they do show the very stimulating nature of coffee.

With the understanding that our climate is warming as a consequence of our too-busy culture, we can see how coffee contributes to our individual and collective pathology. This is particularly true given how readily available coffee is and how normal drinking it might seem. You can go to restaurants of all kinds and get coffee with breakfast, lunch, and dinner. You can go to work and get coffee. You can go to the church supper and get coffee. You can go to an environmental conference and get coffee. You can even go to the gas station or the bank and get coffee.

To provide some context for how much coffee we drink here in the United States, let's look at some numbers. By 2012, Americans consumed a staggering 450 million cups of coffee each day, which translates to about 150 billion cups yearly.[11] Think about that for a moment—*150 billion cups of coffee each year.* That's 150 billion cups of hot, overstimulating coffee in our overstimulated country—150 billion cups of coffee as glaciers are melting, forests are being cut, and methane is bubbling from the ocean floor.

According to the digital magazine *Food Product Design,* eighty-three percent of U.S. adults reported they drank coffee in 2013, and daily coffee consumers remained at sixty-three percent of the adult population.[12] According to the U.S. Census website, the total estimated U.S. population for 2013 was about 316 million people, and a little under thirty percent of those people are under age eighteen, meaning about seventy percent, or about 220 million, are adults.[13] If eighty-three percent of these 220 million adults drink coffee, that leaves about 183 million adult coffee drinkers. Dividing 150 billion cups of coffee by 183 million people shows us *820 cups of coffee per adult drinker each year*—820 cups of coffee as our planet heats and destabilizes and as we overemphasize doing and consuming more.

When so many people drink so much coffee, overstimulation can become the new norm. But just as overemphasizing the Yang in how we live our lives disrupts environmental sustainability, continuously ingesting a hot, stimulating drink disrupts personal well-being. With so many of us drinking so much coffee, it can seem normal to be unable to sit still. We might think it a sign of productivity to feel compelled to go from one thing to the next or to multitask. But in all likelihood, all that these activities and compulsions indicate is heat.

So what can we do other than continuing to be overstimulated by coffee? A good substitute for the heat and damp of coffee is green tea. While green tea contains a moderate amount of caffeine, it moves energy down as well as offering a mild lift. While green tea does enliven the mind and move our Qi, its nature is cooling and drying.[14] As a result, over time its cooling nature can help clear out the heat in our bodies, and its dryness can help drain out the damp that our coffee consumption has produced. There are many high-quality, organic, ethically grown forms of green tea that can be substituted for fair-trade coffee.

For those of us who like a rich drink, a good substitute available from a source that is more local than both coffee and green tea is dandelion root. Yes, it's the same dandelion that grows in your yard, at the park, and in the parking lot. Rather than seeing it as a weed and a nuisance that should be pulled out or poisoned with herbicides, Chinese and Western herbal medicine have long valued dandelion.[15] When simmered over a low heat for a few minutes, it becomes a rich, strong-tasting tea with the pleasant bitterness many of us enjoy in coffee.

In one of the many ironies of our era, we spend a great deal of time and money trying to get rid of a plant that is good medicine for us and, by extension, for the planet. In addition to clearing heat and draining dampness, dandelion is also very effective at clearing toxicity, including the poisons that are applied to try to kill it. A widely used Chinese herbal text has dandelion in the section on "Herbs That Clear Heat and Relieve Toxicity,"[16] and a well-known Western herbal book similarly describes it as promoting detoxification.[17]

Dandelion root can be bought dried or in tea bags, but for those of us interested in living a connected and sustainable life, why not go out and pick your own? As best you can, make sure that the area where you're harvesting it is as far as possible from major sources of pollution such as roads, manufacturing, agriculture, or other people's lawns.

Any time that you can find dandelion, you can harvest the roots. Spring and fall in particular are good times to harvest, as the energy and nutrients of the plant are most concentrated in the roots, while some of the leaves are still available for identification. You can also harvest dandelion in the winter if you live in a place where you can find enough of the plant to locate the roots. In the summer, more of the energy of the plant is above ground in the leaves, stems, and flowers. You can harvest the roots at this time of year as well, though the cooler and darker months are preferable.

Use plant identification books or someone's help to make sure that what you have is dandelion root, and then wash it well, making sure that all the dirt is removed. Next, cut the root into small pieces, about half the size of your fingernails, and place it in on a plate in a warm, dry place away from direct sunlight. Depending on the warmth and humidity, in a week or two the roots should be dry enough to make tea.

For one cup of tea, bring about twelve ounces of water to boil, reduce the heat to a simmer, and add about a tablespoon of dried dandelion. Keeping the lid on the pot, cook for ten to twenty minutes, depending on the potency and taste you desire. Strain it and you have a warm drink that is rich and bitter like coffee, but helps to clear out heat and dampness.[18]

In addition to the cooling effects of dandelion tea for our bodies, being able to harvest it ourselves helps cool the planet as well. Unless you live in a tropical place like Hawaii, the coffee you're drinking is being shipped thousands of miles, which is contributing to climate change. Even if it is being grown organically and traded fairly, coffee has a warming effect on the planet from the greenhouse gases associated with getting it to us. And even though green tea is cooling by nature, it warms the planet, as almost all of it is grown in Asia and shipped thousands of miles. Harvesting and drinking our own dandelion tea helps cool us down and limits greenhouse-gas emissions as well.

Another benefit of dandelion tea as compared with coffee is that making dandelion tea provides an opportunity to connect directly with nature. We might like the jolt of caffeine that coffee provides. Or we may prefer the more subtle lift of green tea. But both pale in comparison with the lasting effects of eating and drinking wild food and medicine that we harvest ourselves. As our overemphasis on the new and the international warms our bodies and our planet, returning to the old and very local of gathering wild plants is part of the medicine we need. Rather than drinking something hot from a far-off place, try something cooling and very local like dandelion. It can be good medicine for you and for the planet.

Just as our assumptions about our lives contribute to global warming, so does what we eat and drink. Just as being busy continuously and constantly having new things fuels the fires of climate change, so does the enormous amount of coffee we drink each year. The amount of coffee we consume in the United States, and the amount of overstimulation it creates within us, is a reflection of the level of climate heat that is rapidly destabilizing the planet. If we're going to address the climate crisis and where it's coming from, it's essential that we understand that climate change is not only happening in the world around us—it's happening within us as well. And coffee is a major source of unsustainable fuel for our internal fire.

In addition to the long list of symptoms we talked about earlier, part of the importance of eliminating coffee is that its overstimulation can also compromise our deeper reserves. In our era of climate change, it is of essential importance for our individual and collective health that we understand that too much stimulation weakens the foundation of our well-being. In particular, the upthrusting nature of coffee can compromise our root strength, including what Chinese medicine calls the jing.

CHAPTER 5

The Water Phase

Oil and Jing

From several thousand years of continuous development, Chinese medicine has developed numerous in-depth, far-reaching perspectives on promoting well-being and treating sickness. Part of what is unique about the Chinese medical tradition is that a great deal of it has been written down, unlike other Indigenous medical traditions that are exclusively oral. We can read today what Chinese acupuncturists and herbalists have been thinking about and debating for millennia. Without these texts, it's possible that a gap in practice and teaching of a generation or two might allow large parts of the tradition to be compromised or lost entirely. We're seeing this now with the diminishment and loss of other traditional medicine systems worldwide, including those of Indigenous peoples here in the United States. Fortunately, not only are their numerous detailed traditional Chinese medical texts still available, but many have been translated into English. As mentioned earlier, one of the oldest and most wide-reaching is the *Nei Jing*, or *Yellow Emperor's Classic of Internal Medicine*. The *Nei Jing* is well-known in Chinese medicine as the original text presenting what are commonly called the Five Elements, transliterated as *wu xing*. When Chinese medical practitioners refer to *wu xing*, they are talking about five phases of movement that create and control everything within us and within nature.[1]

As with the understanding of Yin and Yang, with the Five Phases, the emphasis is on balance. Even in our era of a rapidly warming and destabilizing planet, Yang is not bad. Even with our systemic overemphasis on doing, new, and more, we still need the influence of Yang to maintain our health and the well-being of our country. Similarly, each of the Five Phases is essential to our healing and that of our culture. Going from one extreme to another is not a sign of real change or an indication of lasting progress. As with Yin and Yang, the place of health with the Five Phases occurs from harmony among the different physical, mental, and emotional aspects of our lives.

Just as the appropriateness of the term *Five Elements*[2] is open to discussion, the entire original text associated with this tradition has, not surprisingly, been open to many different interpretations. Thousands of years of continuous application of the ideas in the *Nei Jing* obviously have provided ample opportunity for divergent opinions. The Five Element perspective here in the United States is often associated with the work of J. R. Worsley, who drew on several medical perspectives, both Asian and Western, to create his interpretation of the tradition.[3]

As we talked about in chapter 1, the Five Phases tradition includes a list of correspondences associated with each element. In addition to their physical functions, which are often understood similarly in Chinese and Western medical frameworks, the organs associated with each element are also understood to have emotional and spiritual functions as well. Western medicine, like Western culture more broadly, generally sees a world of separation. While there are emerging exceptions, it generally sees the body as separate from the mind, organs as separate from other organs, and humans as separate from one another and from nature. As discussed earlier, the older Chinese medical views assume interconnection: body, mind, and spirit are connected; organs are interdependent; we human beings are connected to the world around us. And rather than searching for some form of absolute truth—as is common in Western culture and Western medicine—Chinese medicine's inductive reasoning is looking at patterns and tendencies in our lives and in nature.

These assumptions of interconnection and inductive thinking have helped create a very well-developed set of associations in which elements are connected to seasons, weather, emotions, colors, organs, stages of life, and deeper energetic nature. For example, Water is associated with the season of winter, the climate of cold, the emotion of fear, the color blue, the Kidney and Bladder, and the virtues of willpower and wisdom. Water is also associated with the very beginning and very end of life, and is the basis not only of Yin, which it is often used to describe, but of Yang as well.

The Yin and Yang of Water

Water by its nature is yielding; it will easily take any form. If you place water in a small cup, it will take the shape of a cup. If you spill water on the floor, it will naturally flow to the lowest spot on the ground and take the shape of the contour. In this respect, water is humble and an embodiment of Yin: it doesn't insist on being one particular way.

At the same time, water is understood to have incredible power. The increasingly strong and numerous storms we are experiencing from the changes in the climate are an example of this. Watching the water rise below our clinic during Tropical Storm Irene clearly brought this global phenomenon home to me. During the second so-called "hundred-year flood" of that year, it became clear that if the water continued to rise, nothing could stop it. Coupled with water's Yin flexibility and humility is its great Yang power.

This Yin and Yang of the Water phase is housed within the human body in the Kidneys. Located in the lower back, they provide support for the rest of the organs.[4] A good analogy that demonstrates the Kidneys' importance is the foundation of a house. If the foundation is strong, it's possible for the rest of the structure to be in good shape. If it's weak—if it is cracked or decaying—then eventually the rest of the house will be affected. The interior and exterior paint on the house might look nice, there may be new windows and a new roof, but if the foundation isn't solid, the house itself isn't solid. As the Kidney is the basis of the body's Yin and Yang, its condition fundamentally affects the well-being of all the other organs. As organs are responsible for physical, emotional, and spiritual aspects of who we are, the effect of the Kidney reaches from the level of the functioning of the body to the inner depths of our psyche and spirit.

Water is also associated with winter, as that season is both the beginning and end of the year. Although there have been recent dramatic exceptions, winter is the season of cold and dark. Here in the Green Mountains, plants go dormant and animals hibernate or migrate to warmer places. Water freezes, days get shorter, temperatures drop, and life in nature slows down. It's not surprising that winter is associated with the Yin as this cold, darkness, and slowing are all examples of the quality of Yin. While the effects of winter might be less dramatic in more southern places, there is still a similar contraction from the relative decrease in temperature and light.

Fear is associated with Water, as it can be about life and death. As anyone who lives in a place with cold winters knows, even with our modern conveniences, extended subzero temperatures can reduce life to basic issues like warmth and shelter. Living in Vermont and Montana has shown me that a week or two of howling winds at twenty degrees below zero can sharpen the senses and make stoking the wood stove a top priority. The reason for this, of course, is that when the temperature is that cold for that long, our basic need for warmth takes on life-threatening urgency.

As with any emotion, fear can be experienced and expressed in balanced and imbalanced ways. Fear may seem on the surface like a negative emotion, but part of what keeps us feeding wood into the stove in the depth of winter is fear—of discomfort from the cold and even of death. If we were completely without fear of pain or of our own demise, we might not be as motivated to keep the fire going during the frigid weather. The commitment to stoking the fire is part of a basic, healthy survival instinct.

The Wisdom and Power of Jing

Also part of the Water element is wisdom.[5] As the basis of Yin and Yang for all the other organs, the Kidney also houses what Chinese medicine calls jing. Very generally speaking, and in a physical sense, the jing is roughly analogous to the modern understanding of DNA—we get half from each biological parent, who in turn received half of their genetic makeup from each of their parents. In addition to providing the blueprints for our bodies, the jing also includes our unique mental, emotional, and spiritual tendencies.[6]

Our jing holds our reserves of Yin and Yang; at the same time, jing is the wisdom of knowing who we are as unique individuals.[7] Just as each oak seed has the energy and the genetic plan to potentially grow into a towering tree, we each have our own internal jing. Every oak tree looks like an oak tree, but the way each one grows and thereby expresses itself in the world is unique. Similarly, we are all human, but how we express our own version of humanity will be unique based on the inherent uniqueness of our jing.

An oak tree will always be an oak tree. If an oak tree were to have an existential crisis and suddenly want to become a turnip, no amount of emotional angst would create this transformation. An oak tree is an oak tree is an oak tree. But within the confines that define an oak tree, there are countless variations. Some oak trees are short, some are tall, some are wide, some are narrow, some have darker or lighter bark, some grow in the sun and some in the shade. Based on its jing and its environmental influences, the way each oak tree appears in the world is unique. Yes, it is still an oak tree, but no, it is not the same as any other oak tree. Similarly, our jing is what makes us similar as humans as well as inherently unique. *Yes,* we as humans generally look the same, but *no,* no one else can express our unique humanity as we can.

In our overly active, overly stimulated culture, we individually and collectively often rely on our deep reserves of jing to keep going.[8] When we suddenly get a second wind, for example from drinking a hot stimulant like coffee, we are likely uncoiling these concentrated reserves. Coffee in particular helps us override our sense of being tired by stimulating us. But stimulation is not strength; it's heat. And stimulation is like running on fumes rather than running on a source of sustainable fuel; you can do it for a while, but there is inevitably a price to pay.

Chinese medicine emphasizes that our jing is best used for two things. First, it's there to enable us to live a long and healthy life. Our jing helps us understand who we are and have the time to express our unique purpose in the world. Second, our jing is an energetic reserve that we can draw on in times of life-threatening sickness.[9] Like a deep-reaching, long-term, generational savings account, jing is a store from which we can make withdrawals when we really need them to maintain our lives. Not surprisingly, Chinese medicine and internal practices such as Tai Chi and Qi Gong emphasize concentrating this energy and living a balanced life to avoid unnecessarily squandering these finite reserves.

The Wisdom of Burning Oil

To understand the state of our own jing, we need look no further than our current use of oil. Our burning of fossil fuels is clearly an example of our collective lack of wisdom about how to live a healthy and sustainable life. Part of the significance of burning oil is that the oil itself can be understood as the jing of the planet. Just as our jing is described as being our dark, concentrated reserve of Qi, oil is similarly both dark in color and a source of very concentrated energy. With the inductive thinking that the microcosm of our lives is reflected in the macrocosm of the planet, Chinese medicine's understanding of jing holds real relevance not only for personal health, but for the root issues of climate change.

In looking at the development of petroleum's use, the iconic image of the first American wells spouting oil out of the top of their wooden structures is significant on a number of levels. The first, which has been discussed by many authors, is that the discovery and use of oil and the industrial revolution are inextricably intertwined. Without exaggeration, the extraordinary growth of production and consumption that has been industrialization would not have been possible without cheap oil. The concentrated energy that oil provides has been a fundamental force that the United States and the world have used to create the infrastructure of the global economy.[10]

As we face peak oil, the era when fuel was literally flowing out of the ground is over.[11] The oil that still exists is deeper below the surface, in much harder to reach locations, and in much harder to access forms. Once we drilled down a few hundred feet in the open plains of Oklahoma or Texas and thick, black liquid came pouring out. Now we drill miles below the surface in places like the Arctic and the deep ocean. And rather than it being in liquid form, large deposits of oil are mixed with sand and shale.

This change in the location, accessibility, and form of oil obviously has economic and ecological consequences. Drilling at greater depths and in less hospitable locations obviously increases the economic costs and has helped to create ecological catastrophes like the Gulf of Mexico spill in 2010 off the coast of Louisiana. Similarly, extracting semisolid oil out of sand and shale is both economically more expensive and ecologically more destructive. As has been discussed and written about extensively in the last few years, extracting tar sands oil costs a lot more money and has a lot greater environmental impact than accessing the oil that once easily flowed out of the ground. Enormous areas are deforested and dug up to get at oil in tar form, and it is attached to rock and dirt. The intensive extraction process and the chemicals necessary to separate it also contribute to the ecological costs.[12]

In addition to these more discussed economic and ecological issues, changes in the form of oil and our access to it also speak to the condition of the planet itself. If the oil of the planet is indeed analogous to our own jing, then our burning of fossil fuels has not only fundamentally destabilized the climate but also depleted the planet's deep reserves as well. Bill McKibben notes in *Eaarth* that it is no coincidence that peak oil and climate destabilization are occurring simultaneously.[13] From the viewpoint of Chinese medicine's understanding of jing, we can see a clear connection between the two.

Part of what the oil itself has been providing the planet is a source of deep, cooling reserves. While part of the nature of oil is concentrated energy—concentrated Yang—another part of its nature as the planet's jing is its concentrated cooling—concentrated Yin. From this perspective, not only will the burning of fossil fuels increase greenhouse gases, thereby warming the climate, but the loss of the oil reserve itself could have a warming effect.

It's not coincidental that the end of easy-to-get oil is occurring as the climate warms and destabilizes. *The connection between the two is the nature of the oil itself.* Part of the effect of oil as planetary jing is to keep things cool and stable. Oil is a dark, heavy, concentrated fluid, all of which speaks to its Yin nature and its similarity to our own jing. The fact that it takes millions of years to create oil is also analogous to the generational nature of our personal jing. Within us, our jing is passed on generationally. Similarly, we can see oil as coming from the lives and eventual deaths of countless animals and plants whose carbon and Qi were concentrated into black, thick, viscous, energy-concentrated oil.

Of course, we are not simply removing oil from the ground; we obviously have a very specific intent—to burn the oil as fuel. When oil is burned, greenhouse gases are created that trap heat in the atmosphere, warming the planet. But this warming can also be understood to come from the removal of the planet's deep reserves of coolant. The connection between jing and oil implies that climate change is a process that involves more than just heating, but an equally important decrease in coolant. As we convert oil into greenhouse gases that warm the planet, we are simultaneously losing the planetary jing's cooling capacity. Forests, oceans, and glaciers are losing their ability to keep things stable because our burning of the planet's jing is compromising its Yin. This is increasingly important as the processes of increasing heat and decreasing coolant can interact in increasingly dramatic ways.

As mentioned earlier, in chapter 2, senior NASA scientist James Hansen and other climatologists are describing feedback loops due to changes in climate that were once thought to be distant and unlikely possibilities that are now happening. For example, the rapid collapse of Antarctic ice shelves and the melting of Siberian permafrost are happening at much faster rates than were recently anticipated. Simultaneously, huge, unanticipated plumes of the potent greenhouse gas methane are bubbling from the ocean floor. And the oceans themselves may have reached the saturation point of the greenhouse gases they hold. As all of these very significant changes happen simultaneously, the planet is warming quickly and the weather is becoming more unstable.

In understanding oil as the planet's jing, an important question becomes: *What does it mean for us and for the planet itself when there is global jing deficiency?* As is often the case when looking through the lens of Chinese medicine, the question can be answered on many levels. As already discussed, the first and more physical answer is that the climate can warm, especially as the planet's cooling jing is converted into warming greenhouse gases. At a deeper, and perhaps more important level, our use of oil can also be seen as a loss of the collective experience and wisdom of several millions years of life on the planet.

Our personal jing is passed along to us from our parents. From this generational perspective, the jing we have is not ours alone. We've gotten it from our ancestors and will share it with our descendants. At the deepest level, jing is about wisdom. It's what connects us to the countless generations who have come before us and to the generations that we hope will come after us. Our individual jing is central to our individual long-term health, and our collective planetary jing is crucial to long-term environmental sustainability.[14] The planet's jing holds the wisdom and life experience not only of the people who came before us, but of our nonhuman ancestors as well. For millions of years, countless animals and plants have lived, died, decayed, and become oil. We can see all of that living, adaptation, and evolution as contained within the oil itself. The oil is not only the concentrated energy of all of that life, but it's also the concentrated experience and wisdom of those innumerable generations.

Part of what our global economy is doing is converting this ancient planetary jing into the production of disposable things for short-term profit. In light of the information we now have about climate change, both scientific and personal, our use of oil speaks to our individual and collective lack of wisdom.

Cultural Jing Deficiency

Only from a perspective of separation could we ever assume that burning massive amounts of oil might somehow not affect our individual and collective well-being. Think about it for a moment. Oil is discovered in Texas and Oklahoma in the 1800s. We quickly learn to convert it into a fuel usable in transportation and manufacturing. More deposits are found around the world, and between then and now we burn billions and billions of barrels of it.

By the end of the nineteenth century, we did have evidence that burning oil would create gases that would warm the planet. But even more importantly, the holistic thinking of Chinese medicine makes it clear that pulling anything out of the ground on such a huge scale will affect the planet. This is particularly true when we then combust it to create energy. It's only from a perspective that is unable to see things as interconnected that we're unable to anticipate that drilling for oil and burning it on a massive scale will eventually change the planet.

With even a basic understanding of Yin and Yang, we could have anticipated that burning anything on such an enormous scale, year after year, would create warmth. As things overheat, Chinese medicine shows us that the coolant can evaporate, creating Yin-deficient heat. We might not have known specifically that methane gas would be released from the ocean or that the ocean would reach a saturation point. But an understanding of Yin and Yang demonstrates that as the planet warms, its coolant will eventually decrease.

That we didn't see something like climate change coming speaks to our lack of cultural wisdom. You don't need to be a scholar of Chinese medicine to appreciate that continuously pulling things out of the ground will affect the planet. You also don't need to be a climatologist to recognize that burning billions of barrels of concentrated fuel will eventually warm things up.

And the use of the oil itself is directly contributing to this condition as our lack of wisdom is consuming the jing of the planet. We are not only affecting the planet's deep reserves; we're also depleting a source of our own personal and societal wisdom. Everything within us and around us is composed of Qi. With the understanding that microcosmic and macrocosmic events play out the same way on different scales, what is happening to the planet and its oil must be affecting us.

As with the overall state of our climate, there is good news about the condition of our jing. With a country and a culture that lack wisdom, there are well-established methods to strengthen our deep reserves. There are several Chinese herbs that can help do so. *Cuscuta,* which is a dense, black seed—like jing itself—is well-known in Chinese medicine for tonifying our foundational strength. Prepared *Rehmannia* root is heavy and black—again like our jing—and is also recognized to replenish our deep reserves. *Polygonum* root is a dark, blackish-colored root that is dense like *Rehmannia* and is also a well-known jing tonic. Together, they are three well-recognized, much-discussed herbs that have been used for generations to tonify jing.[15] There is also a Chinese herbal formula that combines these three herbs with four others to fortify our deep reserves; it has the poetic English translation of

Seven-Treasure Special Pill for Beautiful Whiskers, and it is also used specifically to tonify jing.[16] Considering the state of the planet and how it reflects our own internal condition, it's not surprising that we prescribe these herbs frequently at our clinic.

As we are all individuals, each with a unique balance of the Five Phases, not everyone will need jing tonics or be able to digest them. As is the nature of jing itself, jing tonics are often heavy, and we need to have enough digestive Qi to assimilate them. Individual herbs and herbal combinations are most effective when they are formulated to address a person's individual condition at a specific time. While many of us could clearly benefit on many levels from jing tonics, we don't all need them all the time. However, it's clear that our country and our culture are in urgent need of generational wisdom right now.

From our usual Western perspective, it might seem like a stretch in logic to equate oil with our jing and the planet's oil with our own health. You might be asking, *How can the oil in the ground really affect my health? And what does our understanding of individual health have to do with our understanding of global sustainability?* But our doubts about the links between the personal and the global, between individual health and climate sustainability, speak to some of the basic beliefs that have helped to create the climate conditions that now confront us. In particular, these doubts speak to our conscious and unconscious belief that we are separate from the world around us.

Rather than questioning our connection, Chinese medicine historically asks, *When could we ever be disconnected from nature?* Rather than wondering if our jing and the deep reserves of the planet are connected, Chinese medicine's holistic view understands that we are a reflection of nature, and nature a reflection of us. For several thousand years, Chinese medicine has featured a deeply embedded understanding that the microcosm of our individual lives is a mirror of the macrocosm of the natural world.

Using the inductive thinking that characterizes Chinese medicine, we can come to recognize that part of the importance of leaving the last of the oil in the ground is that it will reduce greenhouse gas emissions. And, at least as importantly, it will preserve some of the collective experience and wisdom of hundreds of millions of years of life that is the jing of the planet. In this era of short-term profit, pathological materialism, and countless other signs of a culture lacking wisdom, jing is part of the antidote to the sickness of climate change.

CHAPTER 6

The Consequences of Continuous Growth

Wood and Wind

Just as all of our organs are interconnected, our bodies are connected to our minds and emotions, and we are connected to the people and the world around us. When something as foundational as jing is affected within us or on a global scale, it will inevitably affect other things. Since our jing is housed in our Kidneys—those foundational organs—and since our dramatic depletion of global oil reserves mirrors our own collective lack of wisdom, all of the phases and organs will inevitably be affected. The pervasive effects of these fundamental causes are particularly apparent in the transition from Water to Wood, which is the transition from winter into spring.

We experienced this seasonal change here in Vermont as I started writing this chapter. As is happening around the globe, the weather is changing here in the Green Mountain State. A dramatically warm and dry winter had led to an even hotter and drier spring. It's March and it's already eighty degrees Fahrenheit. I'm looking out the window from the desk where I'm writing, and the grass is already green. Going outside, you can already feel the heat radiating from the soil as if it were summer, even though that's three months away. People are out in shorts and T-shirts even though this is often the time we get dramatic late-winter storms that bring bitter cold and feet of snow. These warmer temperatures might feel normal in more southern locales, but here—about an hour's drive from the Canadian border— seeing this much grass and feeling this much warmth at this time of year is disconcerting.

Part of the reason it feels this way now is because we just finished the winter-that-wasn't. We're accustomed to lots of cold here in northern Vermont, where people ski, snowboard, and ice fish for four or five months a year. But this year, rather than freezing temperatures, we experienced several days in January and February when temperatures were in the fifties—cold if you were living in Florida, but startlingly warm for here in the northern Green Mountains. And rather than snow this winter, we've had rain.

Except for two short weeks, the lawn I'm looking at as I type this chapter has been exposed for the whole winter—no snow, no ice. Throughout the winter we experienced overnight lows well above freezing. While Vermonters are accustomed to a January thaw, a "thaw" implies that freezing has occurred—and there was almost none of that this year.

In understanding the connection among the seasons, it makes sense that a hot winter with almost no snow would lead to a hot spring. Part of the nature of the Water phase and winter is that it's the foundation of Yin for the rest of the seasons. The dramatic lack of cold temperatures and almost complete lack of snow this winter indicates that Water's usual cooling and moistening effects are not available to be shared with the Wood element that is spring.

One of movements of the Five Phases cycle is clockwise, from Water to Wood to Fire to Earth to Metal, then back to Water. As part of this particular flow of Qi, one element feeds the next in the cycle. Called the Sheng cycle or Nourishing cycle, it is a supportive relationship that allows each element to feed the one that follows.

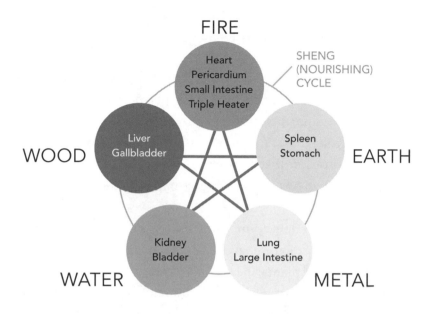

The Sheng Cycle

Rather than being an abstract theory, the Sheng cycle comes from very concrete observations of nature and interactions with the natural world. Taking the relationship between Water and Wood as an example, it's clear that the melting of snow that usually accumulates here in Vermont in winter helps create the growth of spring. We can also see this with houseplants or in our gardens. If you want a plant to grow, it's essential that it gets enough water. Without the moistening, yin effects of Water, the Yang growth that is Wood isn't likely to happen.[1]

As described in the original Five Element text, the *Nei Jing*, within the human body the Wood element is associated with the Liver and the Gallbladder. Wood's season is spring, its climatic factor is wind, its emotion is anger, its sound is shouting, and its stage of life is adolescence. Wood is associated with the eyes, the physical aspects of sight, and the bigger issues of a vision for one's life. Wood is also associated with change.[2]

As one of the foundational traditions in Chinese medicine, the Five Phases provide another deep-reaching perspective into our own health and the sustainability of our culture. It is clear that the United States places a strong overemphasis on Yang. Looking through the Five Phases lens, it's also clear that not only are we depleting the jing of the planet, we are stuck in one of the elements rather than moving healthily from one to the next.

Along with interconnection, another central tenet of Chinese medicine is change. In nature, everything is in a state of dynamism. Animals, insects, trees, plants, the weather, the seasons are all changing continuously. Sometimes the change is dramatic and very noticeable and sometimes it's more subtle, but change is happening all the time. Even rocks, which might appear static, are constantly changing. If you take the time to observe something as seemingly stable as a stone, you'll see that its color and texture are changing. You'll also notice that over time its size and shape change from the effects of the weather, and its color changes from the effects of the sun and rain.

To help us understand where we in the United States are in the Five Phases cycle, let's look at some more vague, open-ended questions. As you did before, ask yourself:

Is expansion better than contraction?
Is shouting for what's right better than being quiet?
Is competition what drives nature?

The Belief in Continuous Growth

As has been discussed by many authors, the ecological effects of an economy based on growth are far-reaching. Despite what advertisements and manufacturers might imply, the things that we buy do not simply materialize out of thin air. Things like shoes, cars, and phones come from somewhere and are made from actual things. And much of the manufacture of these things is unsustainable both in terms of what is being taken from nature as well as what is being returned in terms of waste. We are converting physical aspects of nature into commodities that we can buy and sell, often creating toxicity in the process.

The overwhelming emphasis of our economy is on buying new things and, soon after, replacing them with more new things. As we discussed before, this economic emphasis is on the Yang, and from a Five Phases perspective it's also about the Wood. And as it comes from basic assumptions about the world, this overemphasis on both Yang and Wood can also be seen in many other aspects of U.S. and Western culture. One obvious example is the amount of new information being created. The sheer number of new ideas, new perspectives, and new technologies being disseminated in all aspects of life has reached a fever pitch. There are even new life forms being created through bio engineering. While some of these ideas and technologies are important and worthwhile,[3] the Five Element perspective shows us that newness is only one part of picture. As always, the issue is one of balance, and it's clear that we have become infatuated with the new.

We may hope that we can live healthy lives and have a sustainable world with this strong emphasis on the new. We may think that we can have all of this continuous newness and still live a balanced life and a have harmonious society. But a basic understanding of the Five Element cycle clearly indicates otherwise.

Whenever we overemphasize one aspect of our lives or one aspect of a culture, there are inevitable consequences. Just as our cultural overemphasis on the Yang of new and more has affected the ecological coolant of the planet, our overemphasis on Wood is affecting the other phases. Looking at the Five Elements circle, the Sheng cycle is represented by the line on the outside that moves clockwise. As the image demonstrates, Water feeds Wood. In other words, the growth and new-ness that is the Wood draws its strength from the Water, just as water-ing a plant helps it grow. For something to grow continuously, it will necessarily pull from the reserves that are stored in the Water phase.

On a larger scale, for an economy to grow continuously, it needs to constantly draw from some deep reserve of energy. In the indus-trial era, this reserve has been oil. Without this source of concentrated energy, what's called the progress of industrialization simply would not have been possible. In other words, without a sustained source of Water, there could not be the continuous growth of the Wood. And with our cultural infatuation with continuous economic expansion, it's inevitable that the deep reserves that are the Water will be compro-mised. As the jing of the planet, oil has literally fueled the quest for the growth of the U.S. and global economy for over 150 years.

Rather than a surprise, peak oil is an inevitability. All the growth and the supposed progress that we equate with our modern society had to be fueled by something. The energy to build things, make things, market things, and move things around has to come from somewhere. A large part of this somewhere is below the ground, and the something is oil.

To say that we have been using oil at a faster rate than it can be replenished is a huge understatement. What has taken millions of years to create, we have used up in less than two centuries. In hindsight, once we knew that it takes millions of years to create oil, what did we think was going to happen to the supply? Did we think that we would always be able to find more? Were we hoping that somehow the biological processes that created oil would speed up to create more for us?

We are infatuated with growth, and infatuation requires that the scope of the questions asked and answers sought be limited signifi-cantly. When we take off the glasses that have us seeing continuous growth as possible and desirable, our view of the landscape changes. Rather than a sign of progress and development, the quest for continu-ous growth—seen through the lens of Chinese medicine—is revealed as inevitably pathological and destructive.

Wind in Health and Sickness

As the Wood element is associated with spring, it's about growth, change, and the beginnings of things. When it's in balance, Wood is the buds bursting forth from the willow branches as the early season begins to warm. It's about the spring peeper frogs coming out of their winter dormancy and shouting their iconic high-pitched mating calls in ponds around New England. It's the first appearance of the green grass after the winter snow has melted. It's the dandelion leaves pushing up through the once-frozen soil into the spring sunlight.

With this growth and change in spring comes an increase in wind, which is the weather associated with Wood. Several weeks ago here in Vermont it was obvious that spring had appeared because there was a sustained increase in wind for several days. While the rising temperatures and increasing sunlight also spoke about the arrival of spring, the increase in wind was the clear indication that winter had left.

Part of what wind does from the viewpoint of Chinese medicine is promote change.[4] Wind as an agent of change is evident not only in the changing of seasons, but also in storms, where wind is often generated. One way to look at the importance of the Wood element is that all change—for better and for worse, toward health or sickness, toward sustainability or unsustainability—comes from the influence of wind. Given the relationship between Water and Wood, it's not surprising that we are seeing an increase in storms and strong winds globally.

As we talked about earlier, an exhaustive amount of data indicates that the planet is warming and that there is a corresponding increase in extreme weather. While Western climate science is still looking to make clear and definitive connections between an increase in global temperature and the increase in storms, Chinese medicine can provide some clarity.

With the help of Yin-Yang and the Five Phases, we can see the connection between the increasing temperature and the increase of wind. As discussed, whether it's within us personally or within the planet as a whole, when things get hot it's likely that the coolant will get cooked off. As the water element is the basis of this Yin, these deeper reserves of coolant and stability can become depleted as things heat up. Looking at the relationship between Water and Wood, if the relaxing nature of the Yin of Water is compromised, what is fed to the Wood is hot and agitated. This excess heat can overstimulate the Wood and increase its tendency toward movement, which is an increase in wind. Within our bodies, this increase in wind and movement can create sickness of all kinds, and on a planetary scale it can be seen as an increase in severe weather.

Part of what Water does within us is root things down. When this downward movement is compromised, the naturally upward-rising nature of Wood can become excessive. As a result, we can see a wide variety of issues individually, including symptoms like headaches (especially migraines), high blood pressure, vertigo and dizziness, sleep disturbances, and excessive anger and rage.

This increase of wind on a global scale presents itself more literally as an increase in storms and dramatic, violent weather. This global increase in wind and storms not only makes sense, it could have been predicted for two reasons. First, once we understand that Water is the basis of Yin, we recognize that when the Water phase is hot, everything will eventually heat up. Applying the wisdom of the Five Element model to the climate, we can see that the burning of the planet's jing would warm the climate while depleting the planet's coolant. The deep depletion of the Water element globally makes it likely, if not inevitable, that there will be an increase in wind and storms as the upward movement of the Wood becomes unrooted.

Second, there is a close relationship between Water and Wood as the direct line between the two indicates. The Water feeds the Wood; what happens in the Water affects very directly what happens in the Wood. If the Water is hot and dry, what is passed along to the Wood will be hot and dry. As the planet warms and its coolant decreases— indicating a hot and dry Water element—we can predict a general increase in wind and storms as a likely consequence.

And where does this decrease in the planet's Yin and jing, and the increase in storms and wind, come from? All of it, of course, begins with us. While all of the forces that help to maintain balance in nature are present within us, it is our individual and collective imbalances that are the root issues of the climate crisis. Though we have the same influences internally that have the Yang of day transition to the Yin of night and that move one season to the next, it's our lack of well-being that is the underlying cause of global warming.

The rapid destabilization of the climate, the decrease in the ability of the planet to sequester carbon, the release of potent greenhouse gases from the ocean floor, the burning of the planet's jing, the increase in storms and violent weather—*all of this is a reflection of what's happening within us.*

Not only is our condition connected with that of our natural climate, but the systems and institutions we've created also mirror our imbalances. In particular, our economic system is contributing significantly to climate change, as it is a magnification of our own individual and collective assumptions and imbalances.

Wood Excess and the Yin and Yang of Economics

Trying to maintain something that is out of balance takes a lot of effort. Something that's occurring harmoniously flows along naturally, as would a river moving downstream without many obstacles. Our overemphasis on the Wood has led us to often dramatic measures to maintain growth because the expansion we desire is far beyond anything that can be sustained. One obvious example is our economic policy here in the United States.

In 2008 the much-discussed collapse in the housing market contributed to a dramatic decrease in the size of the economy. In an effort to prevent future decline, the U.S. government infused over $700 billion, including money given to large banks and the auto industry. To some, this was a needed action to prevent an economic collapse. To others it was the federal government bailing out businesses that were on the verge of bankruptcy due to their own greed or poor business practices. Regardless of your particular interpretation of the government bailout, the hundreds of billions of dollars can be seen as a massive, perhaps unprecedented, injection of energy into the economy.

The rationale for this and other attempts to keep things growing can be seen in the words we use to describe the economy. When the economy is "healthy," it is growing at or above a predicted rate. If the growth is slowing or if the economy is not growing at all, this is called a recession. If the economy is contracting, we call it a depression.

For many of us, *recession* and *depression* are scary words. The way these terms are used in both casual public discussions and more in-depth academic discourse speaks to the deep emotion many of us have about the economy. With the fear that these terms create, recession and depression are things to be avoided at almost any cost. We might ask: *Who in their right mind would want a recession or, even worse, a depression?*

The words we use to describe expansion and contraction speaks to what we value. It also reveals how we have structured our country and our culture. As with the earlier question about doing and not doing, it may seem obvious that we should all want the economy to grow. Listening to public radio yesterday, I heard a commentator ask rhetorically, almost word-for-word, the question I posed above. It was clear from his tone that he believed it self-evident that no one in their right mind would, in fact, want the economy to contract.

The economic system we now have is largely based on the belief that it is not only possible but desirable to have something expand continuously, theoretically forever. Not only would Yin-Yang theory say otherwise, but the Five Phases tradition makes it clear that trying to keep something growing all the time is like hoping for it to be spring year-round. Simple observations of nature fundamentally question this assumption. When we assume that anything can continue to grow forever—whether it is a tree, a person or an economy—we are assuming that things will stay in a state of Yang forever. We're also assuming that it will always be spring and never autumn or winter. Just as the Yin-Yang symbol shows that Yang eventually becomes Yin, the Five Phases circle shows us that the spring of Wood becomes the summer of Fire, which becomes the harvest of Earth, which becomes the autumn of Metal.

In the Yin-Yang icon, the peak of Yang is the large amount of white and also where the black of the Yin begins to appear. This part of the circle represents the economy reaching its peak and—simultaneously— beginning to contract. Just as the growth of the Yang, by its very nature, brings in the presence of the Yin, the expansion of the economy leads naturally to its contraction. Similarly, in the Five Phases, the growth of spring and the Wood lead to the transformation to another phase and other seasons.

When it comes to economic policy, we don't seem to understand this basic dynamic of expansion leading to contraction and one season becoming the next. In fact, our response to the words *recession* and *depression* indicates that we fear it. Despite what various people might tell us, continuous growth is simply not possible. Even as we try to pump up the economy, natural transitions still occur. Even with the hundreds of billions of dollars that were injected into the economic system, Yang will eventually become Yin and spring will eventually become summer, late summer, and fall.

I hope I do not sound callous in my application of Chinese medical thinking to economic issues. I am aware that many people were affected by the contraction in the economy that began in 2008 and continues through the writing of this chapter. I am very aware that many people have lost their jobs, businesses, and homes as a result of the recession. I am also aware that these losses have placed real stress on individuals, families, communities, and entire countries. But in our era of climate change, propping up the economy without examining our deep overemphasis on Yang and Wood adds fuel to the climate fire. The pursuit of the "next great thing" to help keep the economy "healthy" is, from another perspective, contributing fundamentally to our own personal imbalances and the warming of our planet.

Looking at economic policy in terms of what we see in nature, we're hoping for it to be warm without cold, day without night, and spring without autumn or winter. And despite dramatic efforts to make this happen, things do become cold, night comes, and the seasons change. Even if we were able to somehow make our collective economic experience be that of a warm, sunny spring day all the time, why would we want to?

The first buds on the willow tree each spring are so exciting because of the dormancy of winter. Seeing the dandelion leaves pushing through the ground in May is hopeful because of the cold and dark of December. One basic tenet for healthy living in Chinese medicine is to follow the lead of the natural world. With an individual, an economy, and a country, when things are in a balanced state, things expand and contract, and are active and inactive. In our era of a rapidly overheating planet, these are not only important transitions, they are inevitable ones.

We might hope to avoid the deeper implications of applying Chinese medical thinking to the causes of climate change. We might be more comfortable with keeping the basic systems that we now have in place and reforming them toward sustainability. We might think that it is simply too much, or too difficult, or too unpragmatic to talk about questioning our overemphasis on continuous growth. However, the severity of climate change is a potentially terminal diagnosis. Whether it's on the personal, cultural, or climate scale, severe symptoms are a strong encouragement to examine the underlying causes of the condition, and make the necessary changes. Climate change not only brings into question the existence of countless species, but also life as we know it. The rapid destabilization of the climate is potentially our wake-up call to look at the deeper causes of the crisis, and our motivation to allow these deeper changes to occur.

A Proportionate Response to Climate Change

While Easterners and Westerners are increasingly integrating one another's beliefs systems into their own lives, the two medical traditions often retain different assumptions about how to treat symptoms. Generally speaking, a more Western view is that symptoms are things to get rid of. If you have a headache, you can take a pill to get rid of the pain. If you have a more serious symptom, you can take a different and stronger pill to make it go away. If that doesn't work, you could have surgery to repair or replace what's causing the problem.[5]

There are times when this approach can be lifesaving. If you are in a car crash or fall off your roof, the emergency room is the place to go. But without looking at where symptoms are coming from, we're not listening to the messages that the symptoms are trying to convey. Of course, there are times when we don't have time to listen to the messenger. If you are having a heart attack, you may not have the opportunity to understand the bigger and deeper issues if dramatic steps aren't taken quickly. Can immediate intervention in severe, acute symptoms be lifesaving? *Yes.* Can fast interventions also discourage us from paying attention to our bodies' messages about our health? *Again, yes.*

Eventually, even if we do succeed in getting rid of the first symptoms, other messengers will inevitably appear. In terms of personal health, it may or may not be the same symptom that reappears. We might get more headaches, or they might go away. They might appear in the same place, or they might move around. But unless we've addressed the root causes, other symptoms will eventually present themselves. With a terminal diagnosis, the general message is that something is out of balance at deep and systemic levels. When a condition can end the life of an individual or the life of an ecosystem, it's a strong encouragement toward change.

Rather than being good or bad things in themselves, serious symptoms are opportunities. In the case of the climate, we can understand a rapidly warming planet as providing us the strong encouragement to look below the surface. We can see a destabilizing climate as providing us the opportunity to look through the symptoms of greenhouse gas emissions and carbon sequestration rates to the more fundamental causes that have brought us where we are today.

In bringing Chinese medicine from the treatment room to the culture, another important idea is what I'll call proportionality. In particular, there is a noticeable lack of connection between the level of climate crisis and the responses being proposed. If someone has the flu but is otherwise healthy, the response can be relatively simple. They might have a mild fever, maybe a slight headache, and perhaps a runny nose and a cough. To help clear things out, the person may receive an acupuncture treatment or two and get some herbs to promote a mild sweat to vent what's causing the symptoms. Their practitioner might suggest some extra rest and sleep, and maybe taking a day or two off work. They might also get some suggestions on soups that would open the pores, and in a few days the person would likely feel better. The sickness is short-lived and not severe, so the proportionate response is straightforward.

If this one case of the flu, however, is part of a pattern of getting sick regularly, the proportionate response should include looking at bigger and deeper issues. The practitioner could ask questions about the person's immune system and their overall strength and vitality. Factors like what they eat, their levels of stress, and their situations at home and work come into play. If someone were getting sick regularly, there would also likely be lifestyle suggestions about diet, sleep, and exercise.

Regularly getting the flu indicates a more long-term chronic condition. An important aspect of Chinese medical thinking is that with the reappearance of sickness, the proportionate response necessarily includes looking at larger issues. Rather than one or two treatments, the person might need weekly acupuncture for a few months. In addition to having an herbal formula on hand for when they get sick, the patient could receive a prescription that would strengthen their Qi while clearing out any viruses or bacteria that might be lingering.

In comparison to getting the flu regularly, a proportionate response to a terminal diagnosis would again require a bigger and deeper perspective. A cancer diagnosis, for example, can obviously bring into question whether someone is going to live or die. There are many different kinds of cancers and an equally wide variety of success rates in treating the condition. Regardless of the specifics, cancer is potentially life-threatening, and the proportionate response would include looking at all aspects of a person's life—the physical, mental, emotional, and even spiritual. This could include not only diet, sleep, and exercise, as with the recurring flus, but also relationships to family and friends, exposure to environmental toxins, expression of emotions, and perspectives about the world.

An increase in the severity of symptoms necessitates a deepening and expansion of the questions being asked, the changes being proposed, and the treatments being offered. A person with a cancer diagnosis might need to come in for acupuncture two to three times weekly for many months. The herbal formula they receive could include very strong substances to clear out toxins, and the quantity of herbs they take daily would likely increase.

Much discussion of climate change noticeably lacks proportionality between the severity of the problem and the proposed responses. Many well written, well-researched writings that demonstrate a clear understanding of the most current science on climate change offer remedies that are disproportionate to the depth and scope of the problem. Most writers propose plans to reduce greenhouse gases while increasing the planet's natural abilities to sequester these gases. There is also a lot of important discussion of eating local food, consuming less, and planting trees. Of course, all of these approaches are essential in treating climate change. But what we eat, what we consume, and the conditions of the forests are mere symptoms of deeper issues.

Perhaps in an effort to seem reasonable and pragmatic, most authors are talking about changes that do little to examine the underlying assumptions that have fueled climate change. The severe global nature of the condition described by climatologists constitutes a potentially terminal diagnosis for life as we know it. The planetary equivalent of taking a few days off work is not an appropriate, proportional response.

Too Much Wind Within Us

As storms increase globally in frequency and severity, it's not surprising that we too are experiencing the signs and symptoms of internal wind. The same force of wind that blows around outside us can create problems within us. Called internal wind in Chinese medicine, it corresponds with the sudden appearance of dramatic and violent symptoms, including migraines, seizures, vertigo, and neurological symptoms of all kinds.

To understand what Chinese medicine means by internal wind, think about wind's strength and speed in nature. Sometimes a light breeze feels good on the skin; it might rustle some leaves or blow around some small, light things you've left outside. Now imagine that the wind is stronger. Branches on trees are starting to move and anything that doesn't have some weight and size is starting to get moved around. Next, imagine a storm in which the treetops are swaying dramatically, branches are being broken, and even bigger, heavier things are being moved by the force of the wind. Finally, imagine that this force that can move treetops and break branches is acting uncontrollably within us, and you understand internal wind.

The good news about symptoms of internal wind is that they are potentially very treatable. One of the more dramatic examples of addressing symptoms of wind I've seen is the case of a farmer I will call Tom. He came to see me because he had been experiencing grand mal seizures about five days a week for several years. Tom was in his mid-forties, wiry, with well-defined taut muscles from decades of regular physical work. He had a sweet smile and a kind and soft manner, but the regularity of the seizures had left him pale and grayish in the face.

During his first appointment, he told me how most days of the week his eyes would roll back into his head as he fell down hard to the ground. He smiled and good-naturedly showed me the scars and bruises all over his body where he had hit the bed, a chair, or the floor as he passed out and the seizures took over. Tom told me how sometimes he would see the change in color that's the aura often associated with a grand mal seizure and have a chance to lie down before the spasms started. Other times it happened so quickly that he didn't have time to respond, and then he hit the ground full force.

Tom's wife described how he shook violently and uncontrollably for up to a minute when he lost consciousness during a seizure, and how afraid she was for him as she watched him convulse. Tom also told me how relaxed but exhausted he felt after the dramatic symptoms had passed. Tom experienced a reduced appetite, which I associated with the stress his whole body was constantly experiencing. His eyes were dull and glossed over, and he had that deer-in-the-headlights look of someone in shock, common with continuous, dramatic, and frightening symptoms. Put simply, he looked frazzled and worn out. Tom's pulse and tongue told the same story as did his symptoms. The pulses underneath my middle finger on his left wrist, corresponding to the Wood element, pounded tautly and were on the very surface of the skin.

Once someone has symptoms as severe as seizures, they have internal wind at the level of a storm. Sometimes the winds are howling—as when Tom experienced seizures—and sometimes they are quiet—as when he felt relaxed afterward—but the presence of too much wind remains constant. Tom's pulses felt as if the Qi of his Liver and Gallbladder were filled to capacity, producing tension like that on the surface of a sail that's being blown before a strong wind. Together with the overstimulation of the wind, the heat in his Liver and Gallbladder was forcing the Qi to the very surface of his skin, creating the pounding quality I felt in his pulse.

Connected to this excess of Wood was a significant deficiency in Water. Along with the significant excess of Qi in the Liver and Gallbladder was an equally significant lack of energy in the Kidney and Bladder. Underneath my ring finger on his left wrist, the Kidney and Bladder pulses were deep and without much force. They were also tight, thin, and pushy. The depth of the pulse indicated that the overall strength of Tom's Water was very weak—in other words, he was Yang deficient. The tightness showed that Tom's Water was bound up and not flowing smoothly, called stagnation. The thinness indicated dryness, as in a river without enough water to fill it out to its banks. The pushiness of the pulse indicated heat, as if the pulse were agitated and unable to relax.

With dramatic symptoms of the kind that Tom experienced, this pulse picture is very common. The excess of wind and Wood can both come from, and contribute to, a deficiency in Water. If the Water lacks enough cooling Yin, what is fed to the Liver and Gallbladder is hot and dry, which can encourage the already rising energy of the Wood to fly up uncontrollably. Also, if the deep strength of the Water—its Yang—is weak, then the ability of the Kidney and Bladder to hold the Wood in place and root it downward is lost. Again, this allows the wind to rush upward. When the Wood and the wind become unrooted and create symptoms like frequent seizures, it can obviously weaken the person at a deep level—namely, the foundational strength contained in the Kidney and the Water.

Tom's tongue told a story very similar to his pulse. His tongue was very red, indicating an excess of heat, and there was a deep crack from the back to the middle, indicating dryness. What was most pronounced was that his tongue deviated dramatically to the side, indicating internal wind.

In addition to the importance of acupuncture and herbal formulas, a very significant part of the success of the treatments came from the changes Tom made in his life. Not surprisingly, my recommendation to Tom was to slow down. A major cause of seizures is the Qi of the Wood quickly and violently rushing upward into the head. An important part of treating too much energy rushing upward is to encourage the Qi to go back down and relax. Being too busy can encourage the already overexcited Qi to become unrooted, triggering a seizure.

While his seizures prevented him from working much, he was still trying to drive his pick-up around the acreage he owned to look at what was growing and how his animals were doing. Even though he could do much less than he had in the past, he was still constantly pushing himself to do as much as he could until the next seizure. During the first appointment, Tom and I talked about a Chinese medicine understanding of what was causing his condition. It all made sense to him and right away he significantly reduced his work on the farm, increased his sleep by several hours each night, and took it easy when he didn't feel well.

After a few months of regular acupuncture treatments, taking herbs daily, and modifying his lifestyle, Tom was feeling much better. His seizures had decreased to about one a week, and the severity of the episodes also decreased. After a few more months, he was having a less dramatic seizure about once a month. Tom also clearly looked better. His face had more color, his appetite had increased, and most days he had much more energy. His face was also much more relaxed; the look of shock he wore when he first came in had been replaced by more sparkle in his eyes.

Comparing Tom's condition to the macrocosm of the climate allows us to see how the increase in wind and storms described by climate science comes from an excess of Qi in the planet's Wood. The planet has warmed, the cooling of the Yin of the Water has been cooked off, and this has created an increase of wind in the Wood. So what's the antidote for this increase of wind and corresponding storms? Just like Tom, we as a planet need to encourage this excess of upward-rising Wood energy to relax and become rooted downward.

As we in the United States and the industrialized West are the source of this wind and storms, an important part of the remedy is to decrease this excessive rising up of Wood energy within us and the culture. Since what we are seeing in the world is a mirror of our own internal condition, we need to break the spell of our infatuation with newness. We need to see clearly this overemphasis on new cars, new phones, new information, new ideas. We also need to recognize that nothing can maintain balance and grow forever. Our often unquestioning faith in new medical procedures and genetically modified life-forms is connected to the storms that are increasing globally. The same continuously upward-rising energy that creates one new thing after the next and that craves continuous expansion is the very same condition that is creating an increase in wind globally.

The Method Is the Message

Looking at the relationship between Water and Wood also makes it clear that we won't be able to shout our way out of the crisis. With the clarity that comes from seeing the root causes, we can recognize the appropriate remedies to treat the cultural issues of our warming planet. It's also essential that we understand the importance of the approaches we use to help inspire change.

The method that we use is indeed an essential part of the message. With our pathological overemphasis on growth, it's not surprising that we see so much excessive anger, shouting, and violence in the world. Just as growth is associated with Wood, all of these other conditions are associated with the Liver and Gallbladder.

We get angry at drivers on the road. We get angry at our family and friends. We regularly hear reports about sometimes lethal anger in our schools and workplaces. In our new-is-better, bigger-is-better, continuous-growth-is-possible culture, a large-scale excess of anger is not only likely but may even be inevitable. An overemphasis on growth in our country helps create an excess of Qi in the Wood element; in our bodies, we experience this as an overstimulation of the Liver and Gall-bladder. As anger is the emotion associated with Wood and shouting is the sound, both of these are, not surprisingly, often in excess in us and in the people around us.

Just as increasing the warmth of Yang in an already hot situation is not likely to bring balance, shouting all the time and being angry about climate change is not likely to create any lasting change. *Yes,* there is a time to shout, but *no,* constantly feeling rage about climate change is no real challenge to the status quo. We might feel strong, even power-ful, when we yell about the state we are in. We might feel righteously justified in screaming about the systems and companies that have contributed to the problem. But being angry constantly and shouting all the time are not likely to create balance in a situation of already extreme Wood excess.

A poignant example of the importance of the methods used to convey our message is a rally I attended in Vermont. It was organized to bring awareness to the proposed pumping of tar sands oil through the northeast corner of the state. About five hundred of us met at a park in downtown Burlington and heard several speakers talk about global warming, other pipeline proposals, and how the issues that affect the environment affect people. Led by a lively, dancing march-ing band with bass drum and tuba, we walked through downtown. Some people sang and danced. Many of us waved to the many hun-dreds of others who lined the sidewalks, looking at what was going on. And as is common at rallies, some people raised their fists and shouted, holding signs with pictures of clenched hands and advocating fighting against various injustices and environmentally irresponsible practices.

When we arrived at another town park about a mile away, we heard more speakers talking about how greed and the role of money in poli-tics are corrupting our country, and how they needed to be addressed to deal with climate change. After the talks finished, several hundred of us changed into black clothing and as a group walked slowly in silence about a quarter mile. We then lay down in front of the hotel where New England governors and eastern Canadian premiers were holding their yearly meeting. We were creating a human oil spill to embody our concern that what had happened with other pipelines could happen in Vermont.

There was no shouting. There was not even any talking. We were not in a rush, and there were no raised fists. What *was* present was the palpable presence of the Yin—the power of the quiet, the peaceful, and the purposeful.

There was a very different feeling, a very different expression of Qi, between the first part of the rally and the second. It's not surprising that with our cultural overemphasis on Yang and Wood that we would assume that being loud was important. We might think that shouting and raising our clenched fists would be the way to get people's attention, make our point, and help create change. But looking at the different parts of the rally through the lenses of both Yin and Yang and the Five Elements, it's clear which has the potential to treat the underlying causes of climate change. In a country suffering from so much anger and violence and straining to maintain an economic system based on continuous expansion, the antidote is not more of the same. The medicine for an excess of Yang and an overemphasis on Wood is the quiet, the peaceful, the energy sinking down and in.

The method we use is not only part of the message we are trying to convey; the method is the message itself.

Another part of how we approach a warming planet is the words we use. They are an expression of our intentions and beliefs, external reflections of our internal condition. Given our overemphasis on growth and newness, it's not surprising that we see things in terms of conflict and war. Medically, we talk about waging war on sicknesses like cancer, diabetes, and heart disease. We talk about our war on drugs and our war on obesity.

When we talk so commonly and even casually about war, we again provide a mirror of our own individual and collective perspective. We may think the wars on various medical diagnoses are not only important but even heroic. But like our emphasis on newness and growth, given the prevalence of anger and violence, looking at health care and social issues in terms of warfare again speaks to our overemphasis on Wood. Similarly, we are not likely to win a war on climate change because in seeing the issue in terms of conflict, we feed the same imbalance and overemphasis that has helped create the condition itself. In our era of Wood excess, a more long-lasting and deep-reaching response is to wage peace.

Biology and Conflict, and the Limits of Innovation

Hiding out in the open, there's another instructive example of our cultural orientation toward conflict: our view of nature. Much of modern biology assumes that the lives of plants and animals are ruled primarily by competition. Looking at the world through that lens, we see everything living on the planet as continuously looking for an advantage. Whether in the pursuit of food, shelter, or reproduction, the assumption is that every organism and every species is constantly looking for the upper hand. Of course, exceptions are discussed, but the overriding perspective is that creatures as small as tadpoles and as big as whales are continuously looking for competitive advantages. Imagining that everyone else and all other species are also always looking for these advantages assumes a state of continuous conflict. Translating this view of nature into a Five Elements perspective, we are again assuming that the Wood element is the dominant reality in the world.

Chinese medicine evolved from cultural perspectives that were rooted in observations of nature and direct experiences with the natural world. The Five Phases tradition does not deny the existence of conflict or dismiss the existence of anger or the potential importance of shouting. Similarly, it doesn't deny that some animals eat other animals or that some plants try to crowd out others. It also doesn't dismiss the reality that in order to sustain our physical lives, we need to eat things that were once living and that may not have given up their lives willingly.

A Five Phases view of conflict, however, is that it's one part of one of the phases. In building our economic system, we have taken one part of the picture and presented it as the whole picture. *Yes,* there is competition in the world, but *no,* the world is not only about conflict. When we see the world as a place about fighting and struggle exclusively, it is both a sign of our imbalance and a contributor to our condition. In particular, it again indicates and feeds our overemphasis on the Wood.

Contemporary Westerners are not the first to see the world primarily as a place of conflict. During the Warring States era, from 481 to 221 BCE, China experienced a seemingly uninterrupted series of wars and conflicts.[6] This included fighting among people and groups within the country as well as warfare between China and other countries. And along with this conflict came significant changes in cultural and medical views.

In medicine, the Warring States era saw a shift away from promoting health and toward combating the evil forces that caused sickness. Culturally, China placed a new emphasis on constantly preparing for warfare among individuals and groups of people. Scholar of Chinese medicine and history Paul Unschuld calls this a perspective of "all against all," in which a war for health and a war against others was inevitable.[7] Modern biology similarly presents this view of continuous conflict in nature as the predominant way to understand the world around us. Rather than among people, as in the Warring States era, modern biological conflict is imagined as occurring among everything in the natural world. We could call it the Warring Nature era of biology.

Like seeing warfare among people as inevitable, viewing competitive advantage as the central force among species also amounts to seeing the world in terms conflict. As I read the history of China from the fifth to the third centuries BCE, I clearly see a time of violence and warfare. And part of what contributed to the unrest was the new ideas that people developed in an attempt to make sense of what was happening in that era. But just as there's nothing preordained about a cultural view of "all against all," there is nothing inevitable about emphasizing competitive advantage in nature. Both are articulations of one particular perspective on human interactions and the interactions in nature. They represent a view of the world as a place where everyone and everything is out for themselves—but this is only one interpretation of reality. In Chinese medicine's view, it's a perspective that overemphasizes one aspect of one of the Five Phases, namely the conflict of the Wood. But continuous conflict is not inevitable; it's one of many possibilities.[8]

Simply seeing our tendency toward viewing the world as a place of continuous conflict and our overemphasis on the Wood won't lead to any significant transformation. And Wood has its value. Just as some animals do eat other animals and some plants do crowd out others, the newness represented by technological innovation certainly has a role to play in addressing climate change. But we won't be able to innovate our way out of the crisis. Innovations in solar panels, windmills, and electric cars are part of the answer to climate change. New, smarter electric grids could also reduce our use of electricity through greater efficiency. But at a deeper level, innovation at best is addressing in the short term the more surface issues of greenhouse gas emissions.

Our often unquestioning faith in newness and innovation that comes from our overemphasis on the Wood is part of our societal pathology that feeds the fire of climate change. A culture that's stuck in Wood will not find the remedy for a rapidly warming planet in more newness. We're also not likely to create any lasting and transformative change by constantly being angry and shouting in the name of addressing climate change. Nor are we going to help ourselves and the planet find balance if we look at climate stabilization in terms of conflict and warfare.

The Need for Vision: The Yin of the Wood

Part of what we lack with our overemphasis on the Yang of the Wood is the Yin of the Wood. The continuous emphasis on new things, conflict, constantly shouting, and always being angry are all examples of the Yang expression of the Liver and Gallbladder. As in any situation, if there's is too much of something, there is likely to be too little of something else. Our overactive Wood has helped to create a very real lack of a clear vision of who we are individually and collectively.

As we discussed at the beginning of this chapter, Wood is associated with the eyes and both the literal and metaphorical aspects of seeing. When the Water phase is dried out from heat, the internal and introspective aspects of Wood that allow us to see ourselves and the world clearly are also dried out. Our preoccupation with new things and new ideas and our emphasis on the loud and on conflict comes at the expense of seeing the consequences of our actions. It can also come at the expense of seeing the underlying causes of our condition. As a result, we are missing the importance of looking at deeper, root causes. Just as our lack of the wisdom of the Water is part of the condition that has created climate change, our lack of a clear vision from the Wood is also part of the collective sickness.

In seeing clearly some of the underlying causes of the climate crisis, we can come to recognize some basic truths. These include that when things are in a state of health, expansion eventually leads to contraction. This is true on the scale of an economy and a culture as well as on the level of an individual person.

We can also come to see that how we approach a crisis like climate change is a reflection of our assumptions about the world. If we view the world as a place of continuous conflict, waging war on climate change might make sense. However, if we're looking to address the underlying causes of the crisis, it's essential that we begin to wage peace.

This movement toward peace includes not only the methods we use to address the many issues of climate change; it also involves how we view the deeper motivations of plants and animals, humans included. More anger, more shouting, and more rage provide us with no real change in our Wood-excess world. Likewise, the belief in continuous economic and ecological competition moves us down the same cultural path that has brought us to the precipice of climate catastrophe.

And just as symptoms like grand mal seizures can be treated, our collective excess of Wood can also be treated. With the wisdom of the Five Phases, we can realize that part of the medicine for the sickness of climate change is in the old and the traditional. Part of the remedy for what ails us and what ails the planet comes from valuing quality over quantity. Part of the medicine we now need includes connecting to something more transcendent than our individual lives.

To treat the excess of Wood in ourselves and our culture that is manifesting as a warming planet, we need to look to the Metal.

CHAPTER 7

Quality Controls Quantity

Metal Controls Wood

If the ideas in Chinese medicine begin to seem abstract, remember: they come from nature. Chinese medicine has many terms and ideas to grasp, but the point of reference is always the same natural world that surrounds all of us. For example, to understand the term *heat,* simply think about standing outside on a hot summer day at noon without any shade. To understand the term *dryness,* think about the soil in your garden that has been exposed to months of hot sun without any rain. To understand the term *internal wind,* think about what it would be like to be outside in a storm, and then imagine that happening inside you.

To understand how the Water phase nourishes the Wood, consider how your houseplants or a farmer's tomatoes need moisture to grow. While this nourishing part of nature is represented by a Sheng cycle line on the outside of the Five Phases circle, the K'o cycle—or the control cycle—is represented on the inside of the circle. While the Sheng cycle describes how different aspects of nature and different parts of our lives feed each other, the K'o cycle indicates how they create limits to maintain balance.[1]

In terms of Yin and Yang, the outside line is about feeding and increasing—which is Yang—and the inside line limits and decreases—which is Yin.[2] Not surprisingly, in the view of Chinese medicine we need to support both the growth and the limiting of each aspect of ourselves and the world. Not only is this essential to our personal well-being, but it is also a vital part of the effective remedy for climate change.

The severe storms that are becoming commonplace are the result of a global excess of the Wood, which mirrors the Wood excess within us. As we discussed in chapter 6, part of what contributes to the dynamic of increased wind is the planet's coolant of Water being compromised. In particular, not only are we creating greenhouse gases that warm the climate, we're also losing the coolant of the planet as we burn the planet's jing as oil. Just as maintaining the balance of Water individually and globally is of utmost importance, it's also essential to recognize the importance of the Metal.

As with the other phases, the Metal has many associations: the Lung and Large Intestine, autumn, dry climates, grief, weeping, old age, tools such as knives and saws, and precious metals like gold and silver. Looking at the Five Phases circle, the Metal is located at the bottom, indicating its descending Yin nature. On a deeper level, the Metal is also associated with precision and our connection to the divine.[3] While it is associated with spirituality and religion, this experience of something greater than our physical selves isn't exclusive to a particular organization or practice. As we'll talk about later in this chapter, not only is this experience of the divine an essential part of our health, it's also essential medicine to treat some of root causes of our warming planet.

To understand the Metal, think about autumn. Here in northern New England, it's the season that follows the harvest, when the abundance of corn, squash, and melons subsides and is replaced with quiet and stillness. The tops of most plants die, and the greens of summer give way to browns and grays that fill our gardens and fields. With autumn comes a decline in energy, and the sunny warmth of summer slowly shifts to a cold darkness.

If you've ever walked through the woods in October and November, you've experienced the Metal. The trees' leaves are dying, and they fall and blanket the forest floor. Though spring and summer are marked by lively sounds, in the fall there is quiet. Walking through a forest of maple and oak, you can only hear the crunch of dry leaves underfoot. Little else in the forest moves around or makes much sound, as animals are getting ready to hibernate or move south. The quiet and stillness of the woods in fall is the embodiment of Metal.

Associations of Metal

Within us, the Lung and Large Intestine are part of the Metal. Similar to the Western view, in Chinese medicine the Lung takes in oxygen and the Large Intestine gets rid of waste. Within a more holistic perspective, however, these organs are more than just their physical functions. The Lung also takes in what Chinese medicine calls the Qi from heaven. Referred to as Heavenly Qi or Ancestral Qi, the Lungs take in this refined energy from the air with each breath we take.[4] Heavenly Qi provides us with inspiration and a direct connection to the sacred. From this viewpoint, heaven is not some far off place that we might get to when we die—instead, we access heaven with every breath.

Divinity is not reserved for particular religious buildings or structures. And it's not exclusive to a particular spiritual sect or creed. It reaches everyone, regardless of their position in society. Equally important, it's not at all confined to humans or even to what we might call sentient life.

This experience of divinity, so essential to our health, fills everyone and everything; there is no hierarchy of sacredness in nature. Just as it's not reserved for people who have a certain amount of money and who look a certain way, it's also not reserved for us humans. All life is sacred—the plants, trees, animals, and insects, they're all sacred. Even things that from our usual Western view aren't alive are sacred. The rocks, the wind, the air, the sunshine, the soil—they're all sacred as well.

Looking back to the physical associations of the Metal, in addition to extracting the last small amount of nourishment from the food we eat and then eradicating physical waste, the Large Intestine also clears out things we don't need mentally or emotionally.[5] Similarly, the Lung takes in the purity of oxygen and Heavenly Qi with each inhale and release impurities with each exhale.[6]

If we look outside our bodies and within nature, we see that the tops of plants in our gardens die off and trees lose their leaves each fall as they are letting go of what's nonessential in order to survive. To make it through the cold and dark of winter, plants concentrate their Qi and their nutrients inward and underground. When it is below freezing for months at a time, trees and plants need to expend only what's necessary; letting go of what's not essential enables this to happen.

In addition to shedding the unnecessary, Metal is also related to precision—think of it as a sharp knife's ability to cut through things cleanly. Here in the northeast, it's important to prune apple trees regularly to concentrate their energy on making fruit rather than making more branches. A sharp saw enables us to make these cuts effectively and efficiently. In order to cut out the excess, the precision and sharpness of Metal is necessary.

Not only must Metal be able to cut precisely, it also is the ability to measure precisely. Being able to measure or quantify something helps us to better understand its value. According to Chinese medicine teacher Thea Elijah, this precision of the Metal is connected to the historical tradition in China of verifying the measurements and weights of scales each year in the fall.[7]

The descending energy of the Metal is also related to the loss of something that was dear to us. It could be something as concrete as a family member or something as open-ended as a stage in one's life. When we have to let go of anything that had real value to us, we are experiencing the energy of the Metal. The emotion associated with autumn is grief, which is one expression of this deep sense of loss, and the sound of the Metal is weeping—another outward expression of letting go.

When we are able to experience grief fully, it opens us up. In particular, it opens the Lungs and can sharpen our understanding of what is truly important in life. In the context of the Metal, this experience is closely connected to the bigger and deeper parts of our lives: spirituality, religion, and questions about who we are and why we're here.

Old age is associated with the Metal as it is the stage of life when, hopefully, we can reflect back on how we've lived and the truths we've learned. Rather than being thought of as a time of failing health and a loss of vitality—as is a common way to think about it in our country—old age is a time when we can contemplate what our lives mean.

Experiencing Metal

To better understand Metal, try the following exercise of sitting quietly with your eyes closed. In a quiet place, make yourself comfortable sitting on the ground with your legs crossed, or sit in a chair with both feet on the floor. Relax your body as much as possible and breathe slow, deep breaths. Lightly place the tip of your tongue on the top, front, center of your mouth, where the roof of your mouth and back of your teeth meet. If possible, breathe through your nose, inhaling and exhaling in a relaxed rhythm. As is done during Tai Chi practice, try to imagine that when you're inhaling you're drawing out a spool of silk, which needs to be done carefully and consistently so as not to break or clump the thread.[8] Follow your breath. Pay attention to its rhythm, note where it's not relaxed or deep, and see if you can gently return it to a steady, smooth pace.

After a few minutes of focusing on your breathing, try to relax your mind. Take note of what ideas or images are appearing, and see if you can let them go. Using your inhale and exhale as encouragement, begin to clear your thoughts. Let yourself stop thinking about your job, or things you have to do, or bills you need to pay. Try not to worry about the next paper you have to write or the fight you just had with a friend. Try not to worry about climate change.

After a few more minutes, open your eyes. Even after a brief period of following our breath and relaxing our mind, many of us will feel more peaceful. Our heart rate has probably slowed and we may feel less stressed than when we started.

The internal relaxation that comes from sitting quietly is part of the Metal. Just as walking through the woods in the fall can help us comprehend how Chinese medicine understands this season, quieting our mind and relaxing our breath can also allow us to experience this phase. The Metal is introspective and peaceful: it's the sense of inspiration that comes with each conscious inhale and exhale of Heavenly Qi.

If you tried to sit still and quiet your mind and had a hard time, you're not alone. Our mental overactivity is at epidemic levels in our overstimulated, planet-warming culture. Similar to how the planet is warming quickly, our minds are overheated. A common result of too much mental and visual stimulation—from computers, cell phones, or TVs, for example—is that our organs heat up internally. This can eventually overstimulate our thoughts, making it difficult to experience peace. Taking away some of the outside distraction, even for a short time, can make it clear—sometimes uncomfortably so—just how constantly busy things are within us.

Even if it's uncomfortable in the beginning, it's vital that we work on cooling ourselves down internally. We can do this by sitting quietly and following our breath or by engaging in more formal sitting meditation. We can also study and practice forms of moving mediation as well. Qi Gong often involves keeping our feet in one spot as we slowly and rhythmically move our bodies and focus on our breath to circulate and concentrate our energy. Tai Chi also helps quiet the mind and focus the energy as we work through a form of movement. In addition to concentrating our Qi and relaxing our thoughts, Tai Chi is also a form of martial arts.

We can also achieve this just by being in a quiet, natural place—like in the woods or near the ocean—and allowing ourselves to relax. Especially for those of us already committed to addressing climate change, experiencing internal peace—where our mind is quiet when we take away the distraction of our busy world—is part of the long-term remedy for our warming planet. It's unlikely that we can help create a more sustainable culture that doesn't continue to overheat the planet if we ourselves are overstimulated.

Meaning Controls Growth: Metal Controls Wood

Given our cultural excess of Wood, in the form of new things and expansion, more Metal is something we greatly need. If something is overgrowing, it's important to not keep feeding it. If you're trying to get rid of the crab grass in your garden, for example, it doesn't make much sense to fertilize it. For us collectively, it also doesn't make sense to keep feeding our belief in continuous growth if we want to address the underlying issues of climate change. What does make sense is to increase the Metal, within us and within our country.

As the line between them indicates, there's a direct connection between Wood (spring) and Metal (autumn). This is not a supportive relationship, however, but rather a relationship of control. If you want a plant to grow, you can ensure this by adding water. If you want to limit and control growth, you must cut it down.

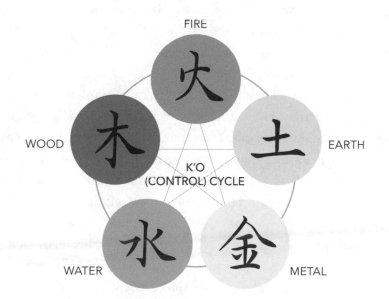

In chapter 6, we talked about how long-term excessive growth in the Wood has been fueled by the Water. In our own selves, this means we're regularly pulling from our deep reserves of jing when we use stimulants to keep going. In our economy, we're drilling and burning oil to maintain an economic system based on a belief in continuous growth. In addition to constantly pulling from the Water, having Wood expand continuously also means that we have weakened the influence of the Metal.

An oak tree doesn't grow to be a thousand feet tall and a tomato plant doesn't take over a two-acre field because there are natural limits. When we chose to create systems based on a belief in continuous growth, we had to limit the Metal's control of the Wood. In the context of our economic system, we're hoping that it will be a sunny, warm spring day every single day. When we're only interested in the spring of Wood and the Yang of warmth and sunshine, we avoid the autumn of Metal and the Yin of cold and dark. Yet, despite our avoidance, just as day becomes night and light becomes dark, spring and summer will always eventually become fall.

Taking into account what we associate with the Metal, if we try to have something grow forever, we're also limiting the natural quieting effects of the spiritual and the religious. Thus, part of the inevitable result of our overemphasis on the growth of the Wood is the loss of the sacred.

Over time, internal practices like sitting quietly and being in nature can help us experience Metal in our everyday lives. Simple things like eating a meal, looking at a sunset, or being someone you love can be of real, deep importance. Deriving such meaning from basic experiences comes from the Metal—it comes from the sacredness in the food we eat, the cycles of nature, and the people we are with.

To be clear, it's not something outside us that enables us to experience the world this way—it's within us, with each breath we take. In Chinese medicine, a connection to the sacred here on earth, and a sense of the divine in everyday life, happens with each inhale. When we're aiming for continuous growth and have whole cultural systems based on it, we're weakening this sense of deeper meaning.

What allows us to erroneously believe that the economy can grow continuously is the dampening of the sacred and the weakening of the limiting effects of the Metal. It would be much harder to continuously and recklessly drill and dig for oil, for example, if we recognized that the ground itself and what's found in it have the same universal Qi that we do. It would also be difficult to continue to cut down forests on a massive scale if we appreciated that the trees are as alive and as conscious as you and I. Continuing to live in a way that destabilizes the climate would seem senseless if we recognized that rather than being connected to the planet, we are the planet itself.

In our era of climate change, spirituality and internal practices are essential aspects of the overall remedy, given their limiting effects on continuous growth. In addition to our own direct experiences with the sacred, listening to ecologically aware spiritual teachers and religious leaders address our overemphasis on the Wood and undervaluing of the Metal is of utmost importance.

In order for something to grow continuously—like the economy, for example—whatever could help control that expansion has to be discarded. The natural limits to growth that we see in the relationship between the Metal and the Wood has to be severed. If we hope that our economic system will grow forever—becoming the monetary equivalent of a thousand-foot-tall oak tree—we have to devalue the relationship between the Metal and the Wood.

An embodied sense of the preciousness of all life and all things in nature is a potent antidote to the quest for never-ending economic expansion. We've had to, necessarily, try to get rid of this sense of the sacred in the quest to ensure ongoing, unchecked economic growth. We can't have the cultural equivalent of a well-sharpened saw cutting away all the things we don't need from our economic system if we hope to have it grow continuously.

In order for the economy to grow forever, the land can't be sacred, the air can't be our connection to heaven, and oil can't be the accumulated wisdom of the planet. For the Wood to continue to expand unchecked in the form of continuous economic growth, trees become resources, people become interchangeable cogs in the industrial system, and food is just another commodity to be bought and sold. When we recognize that trees have the same Qi as you and I, that each of us have a unique contribution to offer the world, and that food is our connection to the Earth, natural and necessary limits are placed on economic growth.

If this loss of meaning and an absence of the sacred seems like a sad state of affairs, it is. We've been persuaded to pursue the next new gadget rather than find real value or meaning in our lives. But ultimately, that new car or phone cannot provide us with any lasting sense of meaning. Not only have we consciously and unconsciously agreed to this deal—of looking for purpose in the pursuit of new and often unneeded things—we call it progress. However, instead of being an indication of advancement, in actuality it's a clear symptom of our sickness.

A new phone, a new car, or a new computer are a hollow replacement for a sense of meaning and a direct experience of the divine. The newness of the world when you open your eyes after meditating, the experience of being filled with vitality after Qi Gong, and the feeling of interconnection from a walk in the woods gives us so much more than the pursuit of the next new thing. A vital part of the antidote to continuous growth is a deep sense of meaning and a direct connection to something bigger than our individual lives.

As we talked about earlier when discussing Yin and Yang, there are advertisers, companies, and stores that are happy to sell us a reprieve from our own internal condition. But just as the discomfort from a lack of Yin isn't treated by buying things, neither is our need for meaning. While new gadgets can distract us for a little while, once that wears off, the basic question remain: What gives our life real meaning?

Meaning Nourishes Wisdom: Metal Nourishes Water

As the Five Phases cycle shows us, in addition to controlling Wood, Metal also supports and helps maintain the well-being of Water.[9] As we discussed earlier, one example of our collective lack of wisdom is continuing to burn oil even though the consequences are clear. One part of treating this is to address the assumptions that foster our belief in continuous growth. Another essential remedy, however, is to understand that a healthy Metal feeds and nourishes Water, both within us and within the planet.

To understand the Sheng cycle relationship between these two phases, think about Water coming out of the ground. Here in Vermont, where many of us rely on private wells, water often contains minerals, as it's filtered down through stone before it's pumped back to the surface. The nutrients it accumulates create what's called hard water, which is nutritious and often cloudy in appearance.

From the view of Chinese medicine, the minerals in well water and the nutrition they provide are part of the Sheng cycle. The hardness of water comes from the minerals that are dissolved, making it more nutritious. In our era of climate change, we need a similar remedy for our own lives and for our country. To help create balance, we need a deep sense of meaning and a direct experience of the sacred that is the Metal to feed the wisdom of the Water to address climate change.

Just as the excess of Wood in our country is depleting Water directly, our belief in continuous growth is also weakening Metal—which depletes Water as well. This is because when we lose a sense of real meaning and a connection to the divine, we also lose what it is that feeds wisdom. A long-term, multigenerational understanding of personal health and ecological sustainability that derives from Water is fed by sitting quietly, moving slowing, and being in nature—all part of the Metal. It's not a coincidence that for millennia Chinese medicine has emphasized practices like meditation, Tai Chi, Qi Gong, and regularly spending time in nature.[10] We need internal quiet, relaxed breathing, and a sense of connection—from internal practices and spending time in nature—so that we can strengthen the Metal, which in turn feeds the Water.

Case Study of Metal Controlling Wood

In our clinic, we regularly witness the Metal's ability to control Wood when treating patients. Mary, a patient in her early thirties who came for help with debilitating migraines, stands out in particular. Mary was fit from exercising regularly, and when she smiled she seemed relaxed. Yet when I asked her about her symptoms, she also had a somewhat faraway look in her eyes—common in someone who experiences severe pain regularly. She also looked tired and pale.

Two or three times a week, Mary would get severe pounding headaches in her temples and at the top of her head. The pain was often accompanied by nausea. The migraines would come on quickly, sometimes taking her by surprise. She could go from feeling fine one moment to needing to lie down in a silent, dark room the next. Mary also had a limited appetite and not much interest in food, even when she didn't have a headache.

Mary's pulse and tongue told a story similar to Tom's in chapter 6. The pulses that correspond to the Wood Element were pounding, and were close to the surface, indicating a severe excess in the Liver and Gallbladder of heat and internal wind. As is commonly the case, Mary's dramatic excess of Wood had caused a significant deficiency in Water. Her pulses that correspond to the Kidney and the Bladder indicated that they were dry and Yin deficient as well as weak and Yang deficient.

Not only was Mary experiencing severe weakness in the Water, she also had a significant weakness in the Metal. Underneath my index finger on her right wrist, there was very little Qi in the pulses of the Large Intestine and Lung. The Qi that was there was much deeper than that of the Wood, indicating that the Lung and Large Intestine were worn out. That both Metal organs were weak was part of the dynamic that had allowed the Wood to become so excessive. As the heat and wind of the Liver and Gallbladder pushed things upward, there was connected tiredness of the descending Qi of the Lung and Large Intestine.

Mary's tongue told a similar story as her pulse. Her whole tongue was very red, indicating excessive heat, and there was a deep crack from the back to the front, indicating significant dryness. As with Tom, what was also prominent was the way her tongue dramatically bent to the side, indicating a great deal of internal wind. At the front of her tongue, Mary also had a wide and deep indent in the position associated with the Lung. As part of the job of Qi is to hold things up, the dip indicated a weakness of the Lung, in particular.

Together, the pulse and tongue diagnoses showed a clear pattern: Mary's symptoms were the result of her Water's inability to cool things down, as well as her weak Metal's inability to control the excess of Wood.

To treat Mary, acupuncture and herbs were used to clear out heat and wind in the Liver and Gallbladder, increasing the coolant and strength of the Kidney and Bladder, and strengthening the Qi of the Metal. To help the Wood relax and to strengthen the Water and Metal, I also suggested that Mary take it easy and not exercise at all until her migraines were better. I encouraged her to rest as much as possible.

After a few weeks of acupuncture, herbs, no exercise, and more sleep, Mary's pain had decreased significantly. When she did have a headache, it was mild and without the dizziness or nausea, and didn't make her lie down in a dark room. After about two months, Mary's symptoms were gone completely, and her energy and appetite also returned. She also had more color in her face, her eyes were clearer, and she looked and felt much better overall.

Quality Controls Quantity

Just as it helped to control Mary's migraines, Metal's ability to balance Wood can also help us on a bigger scale in our country. Because the remedy to our overemphasis on newness and doing is the old and nondoing, we need to increase the calm energy of autumn to check the excess of spring within ourselves and our culture. To help balance the overemphasis on buying more things, we need to appreciate the things we already have. And rather than attempting to create meaning through the pursuit of the next new gadget, we can recognize in nature the sacredness of all life and all things.

Even though we already have so much material stuff in our country, we are constantly encouraged to want more. Not only do many of us have more than we need, much of what we have is designed to be used briefly and then discarded for the next new thing. Because we have more than we need, we value the things we do have less and less.

We can think about this in terms of gardening, too—if you only have one shovel that you use to pull weeds and plant seedlings, it's likely you'll treat it well. If you use it year after year, you're more likely to make sure it's sharp to make digging easier. It's also likely that you'll make sure the handle remains smooth and the wood oiled, so it's less likely to crack or break.

But if, instead of having one shovel, you have fifty, the condition of the one shovel you're currently using probably doesn't matter much to you. Its sharpness or the condition of the handle are less important if you know that if this one breaks, you can just get rid of it. In other words, when you have more shovels than you need, the value of each individual shovel decreases. This relationship between an increase in the number of things we have and the decrease in their value is the connection between the Wood and Metal. The more things we have, the less we value them; this is true for shovels, gadgets, clothes, and nearly everything else.

To help balance our overinfatuation with getting new things, the first step is to truly value the things that we already have. This could be a shovel for gardening, a well-maintained bike to get around town, or the laptop I'm using to write this book. *When we appreciate the things we have, we need fewer things.*

Of course, one reason we continue to buy stuff at a rate that destabilizes the climate is that many products are designed to break or become obsolete soon after they're purchased. Planned obsolescence might increase sales for certain companies, but it's clearly not sustainable ecologically.

If you do need to buy something, try to get the best and most ecological version you can afford. Don't pay for unnecessary frills or something you don't need; invest in quality. For example, you might spend twice as much on a shovel made of hardwood and galvanized metal, but it could last for decades of regular use. With food, buying at a farmers market or directly from a local farm will likely cost more than getting the cheapest meal at a supermarket chain. But the freshness, taste, and nutrition from eating local are part of what you're paying for.

We need to be able to determine the quality of something, which is what the Lung allows us to do. Physically, when we breathe, the Lung knows what to take in—the purity of oxygen—and what to discard—the impurities we exhale. When it comes to objects in our lives, the Lung also helps us recognize what's essential and encourages us to get rid of what's not. Similarly, the Large Intestine helps us extract value out of all aspects of our lives and let go of things we don't need.

If we constantly buy things without much value that aren't made to last, we're overtaxing the Metal. Physiologically, the Lung and Large Intestine must continuously discard the impurities and toxins in what we eat, drink, and breathe, including pesticides, herbicides, preservatives, and a myriad of other chemicals. Outside the body, we're also bombarding ourselves with purchases of meaningless, disposable things we don't need. This not only weakens the Qi of both organs, but also affects our ability to discern what's really important and valuable. In turn, Metal's ability to control the excess of Wood is further compromised.

If we were to change our larger economic system to one not based on the Wood of continuously buying new things, we could increase Metal's influence and place value on repairing and reusing the things we already have. Doing so would also increase the worth of old and traditional tools and technologies rather than unquestioningly having faith in whatever modern convenience is being marketed to us next.

More Metal in the economy would also emphasize and pay for maintenance and restoration—of the land, the water, and the air—rather than destruction. More Metal in our health care system would mean valuing and paying for the promotion of health rather than waiting for sickness. Rather than buying an inexpensive over-the-counter remedy over and over to try to make symptoms go away, you can go to an acupuncturist or Chinese herbalist to address where your symptoms are actually coming from.

To find balance within ourselves and within our culture, it's imperative that we understand that the assumption of excess and more must be balanced with the discernment of reducing and less. It's also imperative that the focus on quantity be tempered with the value of quality. It's also of vital importance that the growth and expansion of Wood be balanced with the discernment and letting-go of Metal.

As we'll talk about next, with the condition of the Wood and Water within ourselves and our culture, other part of our lives will inevitably be affected. Our overemphasis on growth and expansion and our lack of wisdom is affecting the way we communicate—the condition of our Wood and Water is impacting the state of our Fire.

CHAPTER 8

The Dissatisfaction of Too Much Fire

When some part of us or of our country is significantly out of balance, other things will inevitably be affected. As we've talked about in previous chapters, in our era of a rapidly changing climate, the infatuation with continuous growth affects wisdom and our connection to the divine—eventually, too much expansion of Wood will deplete the Water and weaken the Metal.

To treat the root causes of our warming planet, we must tap into the generational wisdom found within our Kidneys as well as into transcendence via our Lungs. In addition, another important part of healing our Wood excess is allowing the Qi to move to the next phase in the cycle, namely the Fire.

To maintain balance, not only do we need the cooling of Water and the controlling influence of Metal, we also need the natural movement from spring to summer. Looking again at the Five Phases cycle, the Qi naturally transitions from the Wood to the Fire if we simply let it. Just as there's no struggle when the winds of spring become the warmth of summer, there doesn't have to be any conflict in allowing ourselves, our country, and our culture to transition from one phase to the next. To heal our warming planet, we must allow things to follow their cycle and change naturally.

As with Water, Wood, and Metal, Fire has a well-developed set of associations. Located at the top of the cycle, Fire is associated with Yang—it is the season of summer, the climate of warmth, the sound of laughter, and the emotions of joy, love, and sadness. The stage of maturity, Fire is also associated with the color red and represents our connection to the people and the world around us. It is also our expression of who we are and is associated with two organs that have Western correlations and two that don't: the Heart and Small Intestine, and the Pericardium and the Triple Heater.

First, let's examine the physical aspects of Fire. The Heart pumps blood and the Small Intestine digests food, extracting nutrients and getting rid of waste. While it isn't recognized as being distinct in Western medicine, the Pericardium is its own organ in Chinese medicine and is the sack around the Heart that also helps with circulation and the Heart's function in general. The Triple Heater is an organ that does not have a clear Western correlation, and is responsible for the movement of fluids and warmth around the body.[1]

Mentally and emotionally, Chinese medicine's appreciation of the Heart requires a more poetic understanding of the organ: the Heart feels and expresses love and joy as well as sadness. The Heart not only pumps our blood, it allows us to express to the world who we truly are. Similar to its physical functions, the Small Intestine is responsible for sorting out the pure from the impure so that the Heart is not affected by mental and emotional toxicity.[2]

The Pericardium protects the Heart mentally and emotionally. Our relationships to small networks and groups of people who surround us are also related to the Pericardium. The Triple Heater helps us regulate our emotional warmth as well as sense the emotional warmth of large groups, such as parties or meetings.[3]

The Warmth of Fire

As I'm writing this chapter in the first week of June, spring has yielded to summer here in northern Vermont. One of the many things that

Vermont's cold winters provide is a real appreciation for warmth and sunshine.

It is about eighty degrees Fahrenheit today—the skies are bright blue with almost no clouds, a rarity in the Green Mountain State, where the regular rain is what makes everything so lush and green. The trees and bushes outside have transitioned from the bright new green of spring to the darker, lusher green of summer. Behind our house, the early summer growth of the maples and oaks is beginning to fill the forest. The ferns that were just pushing through the still cold ground a few months ago are now several feet high and a dark mature shade of green.

Now that summer is arriving, our tomatoes and salad greens are planted in our garden. We've already harvested the first strawberries of the year, and the blackberry and blueberry patches in back of our house are starting to flower. The beautiful purple flowers of wild ground ivy are growing in clumps around our lawn, and dandelions all around our house have already gone from flower to seed.

With the sunshine, warmth, and growth in the woods, there's a lot of fun to be had at this time of year. Where we live, there's music of all kinds being performed in concert halls, on the streets, and in the fields each night. Last night, we walked around town, taking in the various sights and sounds.

As we walked between the concerts, the warmth of Fire was on full display: hundreds of people we sitting outside restaurants and cafés, having dinner, talking, and laughing. Hundreds more were strolling up and down the blocks of shops and restaurants, where there were about a dozen bands playing on the walkway.

On one block, there was an eight-piece ska band cramped onto a small outdoor stage. Down the block, a funk band with a young energetic singer played under a tent. The singer clearly relished the attention and admiration of the dozens of people listening to their music, swaying, bobbing their heads, and dancing. Farther down, you could find a fifteen-piece Latin salsa band, and even more people dancing.

Summer bonfires and barbecues are also starting now, and people head out to the lake to canoe, kayak, and sail. The parking lots at trailheads leading up local mountains fill up on the weekend. Posters have gone up all over town about summer festivals of music, yoga, and local food.

All of the plant growth, warmth, and fun this time of year are that of the Fire. Just as plants respond to the warmth and sunshine, we do too. It's natural and healthy to go with the rhythm of summer: it's the time to visit friends, go to parties, listen to music, be outside, and have a good time. It's a great time to be more active physically and mentally.

When we're in balance with the season, we allow the Yang of the season to lift our energy and our spirits, and increase our activity levels. We can feel our Qi bubbling to the surface in the summer. We feel it especially in our chest as the effects of the Fire move upward into our Heart.

But when there's too much or too little Fire within us, what is normally the joy of summer no longer seems fun. If we have too little Fire, a party doesn't seem worth going to, and we're not that interested in music, bonfires, or barbecues. When we are cold internally, being around other people and having a good time might not seem interesting. It might seem like too much effort, or even impossible, to muster the energy to be outside and be around others.

If we have too much Fire, we are burning too bright and hot all of the time internally. Despite the fact that our country and culture overemphasizes the Yang, having too much Fire is not actually a sign of health. It can be difficult to take in the joy of watching people salsa dancing on a summer night when we are overstimulated. When our organs, thoughts, and emotions are overheated, there isn't enough space for the balanced parts of Fire because the overstimulation has taken over. Even if we are able to partake in the fun of summer, the enjoyment is short lived, and we quickly need to go to the next concert or party because we can't sustain our joy. Rather than creating joy and love, too much Fire eventually creates sadness as the Heart becomes overburdened.

The Dissatisfaction of Too Much Fire

Unfortunately, so many of our attempts to experience the connection that comes from Fire don't seem to give us what we're looking for. Many of the methods we use to communicate give us short-term stimulation rather than long-term warmth. While things like email, texts, and tweets are ways to stay in touch, they don't replace face-to-face and person-to-person interaction. They also can't replace actual time spent in nature—seeing a picture of a beautiful sunset is not the same as the actual experience. Real and meaningful communication is not merely words and images—it's also an exchange of emotions and Qi. Despite what we see in advertisements for cell phones and laptops, an electronic life cannot replace an actual lived life.

The form and duration of electronic communication is, by its nature, short lived and superficial. Not only do the mediums only lend themselves to conveying small bits of information, we've become accustomed to receiving messages in two or three sentences or 140 characters. Some of us receive hundreds of these electronic messages each day, and anything longer than a few words seems like too much.

As we talked about earlier when discussing Metal and Wood, there's a relationship between quantity and quality. Just as the value of things decreases if we have more than we need, our constant sending and receiving of electronic communication can affect the health of our Fire.

More communication doesn't mean better communication. Especially in our already overstimulated world, the quality and meaning of communication is often much more important to our well-being than the sheer quantity of messages we get each day. And when we spend so much time and energy on these electronic messages, it can crowd out more meaningful interactions with people and nature.

In addition to the overstimulation caused by electronic communication, the electricity needed to power our devices also creates heat. Most of our electricity is now derived from unsustainable, planet-warming sources. Even if it is with the best of intentions, to use electricity from coal, oil, and gas is to create greenhouse gases. It's important to recognize that electronic communication is contributing to global warming.

Heat is also created by the very devices we use. Cellphones, computers, laptops, tablets—they all create heat. In looking at the electromagnetic radiation they emit, it's clear that they're another source of overstimulation. Just feel the bottom of a laptop, the top of a hard drive, or the back of a cell phone after using it for a while. It's very likely that it feels warm, even hot. With so many of us using them so frequently and in such close physical contact to the devices when we do so, it makes sense that we're absorbing the heat they're emitting.

The heat from our electronic devices is different from more traditional forms of warmth. Electronically derived heat is more of a buzz of stimulation than the heat that comes from a wood stove. It's more the buzz of Yin-deficient heat than a sustainable form of warmth. There's a frenetic energy to it, where heat and mental activity increase while our sense of peace and internal quiet decrease. The result is a sensation of overstimulation and unrooting, which is a similar dynamic to the effects of coffee that we discussed earlier.

A fundamental understanding of Chinese medicine is that we're strongly affected by our environment. Traditionally, this spoke to the climatic influences of hot and cold and how they could cause internal disease. In our modern era, however, this also includes how the human-created environment affects us as well. From the view of Chinese medicine, when we're in such close proximity to electronic devices that we use for many hours each day, they absolutely impact our well-being—not only in terms of the type of communication they promote or the electricity that's used to power them, but also in the radiation they release. In addition to the stimulating nature of electronic communication, it's likely that the devices themselves are contributing to our overstimulated internal environment.

The Dissatisfaction of an Electronic Life

I imagine many of us have seen the effects of an electronic life replacing a genuine one. One of the most common is to see people, both young and old, looking at their phones constantly when they're with other people. Recently, at a burrito shop I saw a group of high school students who were all looking down at their phones. There was no talking, no laughing, and no one was eating their burritos. None of them were looking at each other and it didn't seem like anyone was having fun. They each had a muted look on their faces as they stared at their phones. After a few minutes, they all looked up, glanced around, and started to eat, remaining as expressionless as they were while looking at their phones.

While we can't always tell by people's expressions what they're feeling internally, it didn't seem as though any of them were really enjoying themselves. All of them seemed to be in their own world, where there didn't seem to be much interest in their surroundings or in their friends sitting next to them.

Of course, there's nothing necessarily unhealthy about being quiet and introspective while eating burritos. But this one, small example is part of a larger pattern around the country. It now seems as though it's more important to communicate electronically with people who are not near us rather than be with the people who are sitting right next to us. In this pattern, experiencing things electronically is more interesting than experiencing the world personally.

Real, lasting warmth and a healthy internal Fire is not sustained by communication in short electronic snippets. These short, superficial methods are both a sign of our cultural condition and a contribution to it.

As we talked about in earlier chapters, we're so overheated and dried out that we're overheating the whole planet and compromising its ability to cool things down. Yin-deficient heat is, not surprisingly, also surfacing in the way we communicate and relate to the world. As the Yin is our sense of satisfaction, what we often lack in the Fire phase is meaningful communication.

Remember the questions from before, and ask yourself:

- *Who am I, really?*
- *How many heartfelt conversations have I had in the last week? The last month? The last year?*
- *Do I feel a real, intimate connection to nature?*

For some of us, these will be easy to answer, but I imagine for many, the responses won't come so readily. Despite the claims that our technological world is bringing us closer together, in many ways it's moving us apart. The act of getting to know someone very well is not a quick process—it's one that requires a physical presence with a person. Likewise, getting to know ourselves intimately and understand who we are and what we're here to do is also not a casual process. It involves clearing out the heat and overstimulation that is a result of living in our current cultural climate. It also requires promoting inner feeling of peace—our Yin—so that we can hear and understand ourselves during a time of cultural Yin deficiency.

This is not to say that emails and texts don't serve a purpose; they can certainly be used to stay in touch when we're apart from our loved ones. However, without physically being with someone, hearing their voice, and seeing their expressions, it's not likely we'll get to know them well. There are exceptions to this, of course, where people have established meaningful relationships with letters or through email. But typically, with our modern, electronic communication, these methods don't lend themselves well to conveying meaning accurately. The smiling and frowning icons that we attach to emails and texts to try to convey emotions are an attempt to express something through a medium that, at best, can only share feelings superficially.

Also, the more we email, text, and tweet, the more we want or even crave them. Superficial communication creates more desire for human connection because, ultimately, it doesn't satisfy us. We then keep communicating superficially, more and more, continuing to look for real connections. But our need for meaningful communication continues to go unmet, and only keeps growing. As is common in our country, these messages claim to offer us what we want and need, rather than give us the thing we need itself—real human connection. Our Fire phase, and our Heart in particular, is fed and sustained through meaningful interactions with people and with the world. Too much superficial communication results in heat and stimulation rather than satisfaction and meaning.

With electronic communication, our belief in more and faster mirrors the Yang overemphasis we've discussed before. In particular, it mirrors the dynamic created by the Yin-deficient heat of consumerism, where buying things gives us a short-term sense of stimulation but little, if any, lasting satisfaction. Because so many of us are hot and lack internal coolant, the way we live doesn't always make it easy for others to take the time to listen and get to know us well.

When we communicate, so many of us are caught up in the Yang of Fire that we lose sight of the Yin of Fire. We eschew the satisfaction, meaning, and depth of deep, sustained communication for a vast number of superficial exchanges. In essence, we've given up the quality of our relationships for the quantity of emails, texts, and tweets. In our era of a rapidly warming planet, this is another branch issue of the deeper causes of climate change.

The Yin of communication means we must slow down ourselves and our minds to be able to hear what someone is really saying. Rather than already thinking about what you're going to say next or what you're going to do after the conversation is over, we must work to be truly present with the other person. Doing so allows us the pleasure of learning about what people are doing and thinking.

Such simple pleasures can be lost in our Yang-excess, overstimulated culture.

However, it is important to remember that in order to really hear someone else, we must first be able to hear ourselves. The Yin of communication with others can only come from the Yin of communication with ourselves. Our overstimulation and lack of satisfaction is contributing to another important issue, one that has reached epidemic levels—we don't know who we are.

To truly know oneself comes from a balance of all of the organs, but it's the Heart in particular that expresses this understanding to the world. Think again about the question: *Who am I, really?* The answer isn't necessarily who we think we are or who we were in the past. It also isn't necessarily who our family or friends think we are or what they think we should do with our lives. Instead, truly understanding ourselves and clearly expressing it comes from a clarity of the Heart. When we know ourselves, we naturally radiate this understanding out into the world through the warmth and light of our inner Fire.

To have this understanding of the self, both the Yin of the Fire and the Yin of the Heart must be present. Without the quiet, peace, and introspection that come from the Yin, it's unlikely that we'll really be able to know ourselves. Not knowing ourselves is ultimately an uncomfortable and unsatisfying way to live.

This discomfort and dissatisfaction might not always be palpable—we might not think about it often or might try to push it out of our thoughts when it does surface. We might feel more comfortable, at least in the short-term, by continuing to do things in the same way and ignoring bigger and deeper questions. We might even take the advice so often given in our overstimulated world: keep busy to avoid the unpleasantness.

But, by itself, the Yang of doing things won't lead us to the Yin of understanding our lives. In fact, in our too-busy world, it often has the opposite effect. Because we lack a Yin understanding of ourselves, we're more likely to be attracted to things that offer us a short reprieve from our discomfort—things like the consumerism of buying things we don't need, the short-term excitement of electronic communication, or the stimulation of coffee.

In looking at climate change through the lens of the Fire phase, it is urgently clear that we need to slow down and do less so we can begin to hear who we are. That we've lost a clear connection to our Hearts is both a cause and a symptom of climate change. Despite the fact that the science is clear about what's happening to our planet, we continue to live in ways that fuel climate change. Not only does this continue to cause deforestation, melting ice caps, methane plumes from the ocean, and a warming planet, it continues to overheat us internally. This overstimulation affects the meaning and connections in our lives, leaving us unsatisfied and looking for the next short-term reprieve from the effects of our internal condition.

Knowing ourselves, knowing what we are here to contribute, and knowing our connection to nature is essential medicine for treating the causes of climate change.

Lack of Vision Affects the Fire

What is currently happening with the Fire shouldn't be a surprise: with our often unquestioning belief in newness and growth, and our tendency toward conflict, we have been severely overemphasizing one of the five phases.

As we can see in the Five Phases circle, there's a direct relationship between Wood and Fire. If you're having a bonfire, a lot of wood means the fire's flames will be large and bright. As the wood gets burned down, however, the flames start to lessen. When there's not enough wood left to fuel the fire, it starts to go out. Similar to a bonfire, our well-being and that of our culture needs a regular, consistent supply of Wood to keep the Fire in balance. If we have too much Wood, the Fire can't burn steadily; the flames will get too large, making the fire too hot, and the fuel will burn too quickly. If there's not enough Wood, the Fire won't burn bright enough, and eventually it will burn out, leading to cold and darkness. Neither of these extremes are healthy or sustainable; the place of health is in the middle.

Both individually and collectively, a notable consequence of our excess Wood is its effect on Fire. As we covered earlier, the Sheng or Nourishing cycle is what allows organs and phases to feed what comes next. While the K'o cycle creates balance by limiting and controlling things, the Sheng cycle is the relationship among the different aspects of who we are that promotes growth—physically, mentally, emotionally, and spiritually. And just as this nourishing dynamic appears within us, it occurs within all of nature as well. Just as the nutrients from Metal feed Water, and the moisture of Water feeds Wood, the Wood becomes the fuel for the Fire.

Given the condition of our Wood, the fuel currently being fed to the Fire is already overstimulated and dried out. This is of real significance, as the Fire is synonymous with Yang and is hot by nature, and therefore prone toward overheating. Our desire for new things and our tendency toward conflict is rapidly fueling the Fire. All of our Wood excess—our already overheated, dried-out economic and emotional fuel—is what we're burning to fuel our relationships with each other and with nature. Knowing this unhealthy, unbalanced fuel is what we're currently using to feed our relationships, it's no surprise that the Fire is burning too hot and too quickly.

This overabundance of the Yang of the Wood equates to a lack of Yin of the Wood—a lack of vision of who we are, both individually and collectively. Without clarity, we lack the right fuel—a Wood that is in balance—to maintain the health of the Fire.

A Lack of Wisdom Affects the Fire: Heart-Kidney Communication

How the condition of our Wood affects the Fire is even further magnified by the current state of the Water. As noted earlier, the planet's coolant has been compromised. On an individual level, the state of our Water is seen in a lack of wisdom—especially as we continue to drill for increasingly inaccessible oil even though science makes clear the consequences. Just as the Wood feeds the Fire, the Water has the potential to control the Fire through the K'o cycle.

When things are in balance, the Yin of Water helps to limit the Yang of Fire. The Water can help keep things cool and prevent the Fire from burning too hot or getting out of control. When you're done enjoying your bonfire, you can use water to put out the flames. But this simple and fundamental relationship between Yin and Yang, between Water and Fire, has been compromised because of how we live and how we see the world. Because we're so focused on Yang, we've compromised the quality and effectiveness of our Yin—which is based in the Water phase and in our Kidneys. When our Kidneys become hot, they lose their ability to cool and control the Fire within us. Our continuous quest for new things and growth—*an excess of Wood*—has compromised our wisdom—*the Yin of the Water*—which is affecting our relationship to ourselves and others—*an overstimulated Fire*.

The Fire's relationship to the Water is a critically fundamental issue. Often referred to as the Heart-Kidney axis, it's considered by some scholars and practitioners to be *the* essential focus of all of Chinese medicine. The axis between Fire and Water is the alignment of who we truly are—the jing of our Kidneys—with our expression of ourselves in the world through the spirit of our Heart, called the *shen*.[4] For some, based on thousands of years of practice and development, the purpose of Chinese medicine is to ensure the health of the Heart and Kidney as well as the communication between them. This allows us to know ourselves and express it clearly. From this viewpoint, we can't be healthy if we don't understand ourselves and can't manifest it to the world.

As we talked about during the discussion of climate change science, the planet is currently experiencing Yin-deficient heat as its ability to sequester greenhouse gas emissions decreases and the climate warms. This is also partially due to the nature of oil, wherein the jing of the planet is being drilled and dug out of the ground. As we noted in chapter 5, the negative effects on our planet are mirrored in our continued use of oil even though the consequences are clear. Associated with the Kidney and the Water phase, our lack of wisdom is on clear display in how we continue to live and view the world.

The overstimulated culture we now live in is mirrored by the warming natural environment around us. The excess heat and excess Yang, as well as the lack of coolant and the lack of jing, inevitably affects the Fire. Not only does the Yin of Water help control the Yang of Fire through the K'o cycle, it is essential for us to be able to express who we are through our Heart.

Just as we lack the wisdom of jing in our own lives and in our culture, we're converting oil—the jing of the planet—into short-term profit. It's not surprising that the methods of communication we've created also mimic this dynamic, where the lack of Yin and decrease in Water create an excess stimulation of Fire. That we've become so accustomed to quickly sending and receiving short bits of electronic information, and that so many of us rely on it as a primary form of communication, speaks to our collective condition. And this condition comes from how we see the world and what we value.

So much of our communication is fast and superficial, and so much of our experience of the world is electronic and removed. Given the condition of our Water and that of the planet, it's easy to understand why so many of us don't really know who we are. Because we don't know ourselves, we can't express our own unique contributions to the world. Based on over ten years of clinical experience treating a wide variety of people with a vast variety of symptoms, it's clear to me that our lack of communication between the Heart and Kidney is at epidemic levels.

Of course, if you don't have the basic needs for survival met, it's difficult to worry about things like who you are and what you're here to do. Without enough food and clothing or adequate shelter and health care, these bigger and deeper questions don't mean much because you're just trying to survive. But if we do have enough of these things— or more than enough, as is often the case in the United States—continual acquisition of more things doesn't create more meaning or more health. Instead, there is an inverse relationship, where more things creates less meaning. Once our basic needs are met, what truly sustains us is meaning and purpose.

If you feel like you aren't sure if you clearly know who you are or can't fully express yourself to the world, you're not alone. Another way to understand the condition of the climate is that our lack of understanding of who we are individually and collectively has reached such an epidemic level that it's affecting the whole planet. We have so lost ourselves in the mire of consumerism, newness, growth, and conflict that our planet may be at a tipping point. The dramatic changes we're seeing globally mirror the depth of our own lack of understanding of ourselves.

The Remedy for Too Much Stimulated Fire

With the understanding that personal symptoms or climate symptoms are messengers trying to lead us back to health, let's talk about some of the remedies for our Fire excess. One simple remedy that we could all do is to simply unplug—shut down your computer and cell phone and disconnect from the Internet. Do it for an hour, a day, a week, or a month. Do it for even longer if you can.

With how overstimulated we are, I can imagine some of your responses. Maybe you think being offline isn't practical because there are messages to answer or that turning off your cell phone prevents you from keeping in touch with work, family, and friends. But if we're serious about addressing the root causes of climate change, we need to be willing to change.

Change doesn't merely mean political changes—ones that help bring more solar panels to our rooftops and more electric cars to our roads. It isn't just economic changes, either—ones that recognize the folly of the quest for continuous growth. True change also includes understanding that our rapidly warming climate is mirroring our rapidly warming internal environment. It also includes realizing that climate change is not only happening outside us but is happening within us as well. Being willing to unplug from hot, fast, and stimulating forms of communication is an importance part of cooling ourselves.

Another reason we're resistant to unplugging is that it can be uncomfortable, at least in the short-term. We live in a Yin-deficient country that's part of a Yin-deficient culture that uses Yin-deficient methods to communicate. Taking away the opportunity to constantly write emails or always answer the phone every time it rings leaves us with increased opportunities to be with ourselves. Because the way we currently live compromises our Yin—which is our sense of peace and satisfaction—it can feel unpleasant to just *be*. Without electronic distractions, we're left with a clearer experience of our internal, unbalanced condition. While it might be unpleasant initially, it's vitally important to our health and the health of the climate that we be willing to experience what's happening within us internally.

Just as there's no other planet to go to if we make this one unlivable, there's no other place for us to reside other than our own bodies. If it's uncomfortable without the stimulation of our electronic gadgets, acknowledge this. If a primary way you communicate with others is electronically, acknowledge this as well.

As is true at our clinic, where each acupuncture session and herbal formula is customized for each patient every appointment, we're going to have to answer for ourselves how much electronic communication we use. How quickly we respond to emails and how many times we check our phones each day likely depends on responsibilities and expectations we have at home, school, and work. Just as there isn't one set of acupuncture points or one combination of herbs that will address a particular diagnosis, the amount of time and energy we should direct toward electronic communication will also vary.

But it is of vital importance to recognize that more of the same will not create different results. Just as the climate isn't going to cool if we continue to increase greenhouse gas emissions, we're not going to clear our internal heat if we continue down the path so many of us are now on. Even if they can be used skillfully to spread the word about what's happening to the climate, more emails, texts, and tweets are not the long-term remedy for what ails us or the planet. In our era of individual and cultural overstimulation, it's likely that many of us who use electronic communication would benefit from using it far less.

With less electronic communication, there's more time and more energy for actual, satisfying communication. More time to say hello to our neighbors and friends. More time to go hear live music. More time to be outside with plants and trees. More time to spend with ourselves. In our era of climate, less is often more.

And just as our overstimulation is affecting how we communicate with each other, it's also affecting us at deeper levels as well. Just as climate change is impacting all parts of the planet, our levels of excess heat internally are affecting us in equally systemic ways. As we'll talk about next, the extraordinary rates of cancer diagnoses in our country are mirroring the rapid warming and destabilizing of the planet. Again, the small picture of what is happening in our bodies is a reflection of the big picture of what is happening in the world.

CHAPTER 9

Cancer and Climate Change

A Tale of Two Diagnoses

According to the American Cancer Society, for those of us now living in the United States, over 120 million of us will be diagnosed with cancer. And over 70 million of us are projected to die from the condition.[1] When confronted with such staggeringly high numbers of such a potentially terminal diagnosis like cancer, basic issues of survival can surface. As we discussed earlier, life and death issues are related to the Kidney, which is also associated with the emotion of fear. When something brings our own mortality into question, it affects the Kidney, as it's the foundation of our lives and can lead to fear and distress. However, if we take a closer look and begin to notice the patterns that are the root causes of sickness, they can become less scary. Diagnoses of all kinds—both Western and Eastern and on small and large scales—begin to make sense as the result of cause and effect.

Cancer diagnoses are primarily discussed and treated from a Western medicine perspective in the United States. As noted earlier, Western medicine is based on the assumptions of Western culture, which often sees the world as a place of separation and conflict. As a result, it perceives organs as separate from each other, with the physical aspects of our lives being distinct from the mental and emotional aspects. It also assumes much of the time that if there's a problem, fighting is often a good response. Such a response makes it difficult to uncover the internal patterns that created the conditions for a diagnosis like cancer.

With the emphasis on compartmentalization and specialization that is common in Western medicine, the focus is on waiting to treat conditions like cancer until after they've already appeared. Western medicine does acknowledge the connection between cancer and things such as lifestyle and exposure to toxins, but it is in general terms and not focused on the individual. However, what we eat, the amount of exercise we get, and our susceptibility to carcinogens certainly affects our likelihood of developing cancer.

Unlike the general blanket statements that Western medicine delivers on what it considers healthy behaviors and healthy foods, the Eastern perspective is more fine-tuned and specific to each individual. Based on the understanding that we are unique individuals with a unique balance of Yin-Yang and the Five Phases, what can be helpful for one person might not be for someone else, or could even create health problems in others. In addition to the importance of lifestyle and diet suggestions being person-specific, there are bigger, more systemic issues that are creating the conditions within us and within our country that contribute to cancer diagnoses. As we'll talk about next, the underlying cultural causes of climate change are similar to the root imbalances that create the proliferation of unhealthy cells within us.

The Roots of Cancer

In looking at our lives, our culture, and the condition of the climate from the view of Yin and Yang, things are rapidly overheating and coolant is decreasing—too much Yang and too little Yin. From the perspective of the Five Phases, we have a dramatic overemphasis on newness, expansion, and conflict, which is affecting our experience of the sacred, our wisdom, and the way we communicate—an excess of Wood is creating a lack of Metal and Water and an excess of Fire.

Not only are these dynamics showing up in all aspects of our lives—from what we value, to what we buy and drink, to the stories we tell ourselves, to how we communicate—they are manifesting in our health as well. With the inductive and holistic thinking of Chinese medicine, this isn't much of a surprise. What we think, what we believe, what we ingest, and how we live all affect our Qi. And the condition of our Qi is the condition of our lives—physically, mentally, emotionally, and spiritually.

Just as the planet is warming as its ability to sequester greenhouse gases decreases, we're experiencing Yin-deficient heat. And just as our economy and culture promote continuous growth, we're experiencing excess within us as well. This heat, lack of Yin, and excess growth are a good place to start the discussion about the Chinese medicine understanding of cancer.

A Western medicine discussion often focuses on the growth of cells. While there are many different kinds of cancer that present in many different ways, in discussing the basics, the American Cancer Society says, "All cancers start because abnormal cells grow out of control." In comparison, "Normal cells grow, divide to make new cells, and die in an orderly way." In describing how cancer spreads, "Cancer cells often travel to other parts of the body where they can grow and form new tumors. Over time, the tumors replace normal tissue, crowd it, or push it aside," creating a metastasis.[2]

As we've been discussing with the condition of our planet, a Chinese medicine understanding of symptoms, both large and small, is that they often originate from deeper causes. Thus, the growth of our cells is affected by the condition of our internal environment.

From my clinical experience treating people with cancer, there's a clear pattern. While everyone is different, and different cancers present differently at various stages of progression, there are underlying similarities. Looking at the Western diagnosis from an Eastern perspective, cells growing out of control fits well with the Chinese medicine understanding of heat. Heat overstimulates us, which can create an excess of growth at a cellular level. As an excess of Yang, this heat propels cells to travel within the body. As tumors grow in size and number, this is again a sign of too much stimulation and too much heat. From a Five Phases perspective, this growth most clearly corresponds with too much Wood. As we discussed, Wood relates to expansion, the climatic influence of wind, and movement. When Wood is in excess, things can start to grow out of control, and when the accompanying wind is also in excess, these things can start to move around the body.

Though we are all individuals with a unique internal environment, all the patients I've worked with who had a cancer diagnosis have had significant amounts of heat. Just as occurs within the environment, within us, if things heat up, coolant gets cooked off. The pulse and tongue of those I've treated clearly indicate this Yin deficiency—and our Yin is what helps to hold our imbalances in check. When we start to lose our Yin, symptoms begin to surface, including very hot conditions like cancer.

Many of us live overstimulated lives within a hot, Yin-deficient country and culture that values Yang and overemphasizes Wood. To understand just how hot, Yin-deficient, and Wood-focused we've become, let's take a look at the numbers.

The Response to the Statistics of Cancer

Looking at information provided by the American Cancer Society, here are some statistics that tell the story of cancer within the United States.

- 1 in 2 men will be diagnosed with cancer in their lifetime.

- 1 in 3 women will be diagnosed with cancer in their life-time.[3]

Together, these numbers indicate that about 2 in 5 people in the United States—or about forty percent—will have a cancer diagnosis. Think about that for a moment. Think about five people you know: five friends, five family members, five people from work or school. If the current trend continues, on average, two of them will be diagnosed with cancer at some point.

With the population in the United States currently at about 320 million, that equals over 125 million cancer diagnoses. *That's 125,000,000 diagnoses of cancer in the United States alone.* This is not 125 million cases of the common cold or flu. Nor is it 125 million diagnoses of frequent headaches. *This is 125 million cases of a severe condition that from a Western view can be life threatening, and from a Chinese view indicates advanced heat.*

Not only are the chances of a cancer diagnoses staggering, but the associated mortality rates are also sobering.

- 1 in 4 men will die from cancer.
- 1 in 5 women will die from cancer.[4]

The simple math is that between twenty and twenty-five percent of the people in the United States are projected to be diagnosed with cancer. Taking 22.5 percent as the average, and with an approximately equal population of men and women, that indicates that seventy-two million people will die from cancer. *72,000,000 people.* So many of us know multiple people who have been diagnosed with cancer—a neighbor, a coworker, a friend, a family member, a spouse. From the above numbers, it's likely even that many of you reading this have been diagnosed with cancer.

Because of how common cancer is, it can take on a sense of inevitability. It might start to seem that people just get cancer, and there's little, if anything, that we can really do about it. We may also be fearful, as it seems that the possibility of cancer is lurking all around us, and that anyone can get the diagnosis at any time.

Just as there are many different ways to respond to our rapidly warming planet, there are many ways—personally and collectively—to respond to cancer. A very common response that is part of the cultural narrative is that we must fight cancer. This war on cancer may be well-intentioned, because we're trying to rally support for people with the condition. Western treatments based on this fighting approach can also sometimes reduce or even eliminate the cells that are causing the symptoms. However, there are often negative side effects as well. As we'll discuss later, this includes treatments that sometimes create the very conditions they're intended to treat.

If we choose to wage war on people's cells, there is often very significant collateral damage. These cells are ones that are needed to continue to live: this war on our cells can affect the health of our digestive system, our strength, our overall well-being, and our will to live. In line with our emphasis on Wood and conflict, the typical Western approach to cancer includes using very strong and very toxic substances to kill cells and procedures to remove the affected areas.

On a bigger and deeper level than just the cellular, cancer also mirrors how we're living our lives. Certainly, there are physical and environmental factors that increase our likelihood of developing cancer, and Western medicine points to the use of tobacco, eating too much meat, drinking too much alcohol, and exposure to a wide range of carcinogens. It's also noted that exercising regularly and not being overweight can limit the likelihood of cancer.[5]

But from an Eastern view, too much of some things and not enough of others increases the likelihood of internal imbalance. Over time, this can lead to conditions like cancer. For example, alcohol is widely acknowledged to be warming or hot in Chinese medicine. Widely published Chinese medicine author Bob Flaws describes it as warm and toxic, both of which could contribute to a cancer diagnosis.[6]

Internationally recognized practitioner and author Giovani Maciocia writes that tobacco was introduced to China in the late sixteenth century, during the Ming dynasty, and that a Chinese herbal text from that era describes it as hot and toxic without any medicinal effects. A Chinese physician, Qu Ci Shan, from the subsequent era of Chinese history, the Qing dynasty, also described how tobacco is hot and drying, and as a result, it burns the jing and fluids, damages the throat,

Stomach, and Lungs, and affects the Heart. Another physician from the Qing dynasty, Zhao Xue Min, observed that smoking also affects the blood, and in general shortened life.[7]

This discussion about the effects of tobacco is interesting and relevant on several levels.[8] The first is that it predates our current understanding of the health effects of tobacco by several hundred years. It's not just a modern Western idea that tobacco can contribute to cancer and decrease our life span. Second, in addition to acknowledging that smoking affects the Lungs, Chinese medicine also recognizes that it can affect the jing. As discussed previously, we live in a time where we're drilling and burning oil on a massive scale. In addition to speaking to our own lack of wisdom that comes from our jing, this process is also consuming the jing of the planet. The use of tobacco can burn out our jing as well—which is very relevant information during our era of climate change.[9]

In addition to the temperature of alcohol and tobacco, in Chinese nutrition some meat is also considered warming. Flaws describes beef, duck, and pork as neutral in temperature, but lamb, venison, and chicken as warming.[10] An overabundance of the hotter meats in particular can overstimulate our organs and our Qi, contributing to an overgrowth of unhealthy cells and potentially creating cancerous conditions.

Waiting for the Problem to Appear

Although it's not commonly noted as a contributor to cancer from a Western perspective, Chinese medicine indicates clearly that coffee also strongly contributes to heat, overstimulation, and growth.

The United States consumes 150 billion cups of coffee a year, contributing to—or even causing—things like anxiety, heart palpitation, and migraines. On a deeper level, coffee can also create heat that overstimulates cells, which can cause them to grow in unhealthy ways and move around the body.

While there is real value in a reciprocal discussion between Eastern and Western medicine, the holistic thinking of Chinese medicine means that we don't necessarily need to wait for definitive proof to makes choices that promote health. For example, we don't need studies similar to those that demonstrate the connection between cancer and tobacco to recognize foods that contribute to the overgrowth of unhealthy cells.

For several hundred years, Eastern physicians and scholars have understood and written about the heat associated with coffee. In Chinese medicine, we see the pattern that contributes to the root cause of the growth and spread of unhealthy cells: heat. It is a significant issue with both our health and that of the planet that Western science and medicine wait for the problem to occur before studying and attempting to address the condition.

This might seem reasonable, necessary, and even prudent from our more usual Western view. The rationale of our culture is often that we need a certain type of data and evidence before we can act. But this deductive reasoning—which had us wait until there was undeniable evidence that smoking causes cancer and that burning oil warms the planet—is actually a part of our cultural condition. In particular, it speaks to a worldview based on separation, in which we ignore the larger picture in favor of focusing on issues in isolation. As a result, we're less able to see the patterns that contribute to forty percent of Americans potentially having a cancer diagnosis and the climate destabilizing rapidly.

Waiting for definitive proof of climate change means that there are already enough greenhouse gases in the atmosphere to dramatically affect the weather. Waiting for definite proof that tobacco causes cancer means that millions of people have already developed the condition from smoking. Rather than waiting for the imbalance to occur on a large scale, it's much more insightful to treat the condition before it's a problem.

This is true not only with the heat of cancer but with the heat of climate change as well. We could have anticipated that burning oil on a massive scale would warm the planet, and likewise, we can also see that within an overstimulated culture, regularly drinking something stimulating like alcohol and inhaling a hot substance like tobacco would contribute to internal heat. We can see this same pattern in the effects of coffee.

To be clear, this is not to say definitely that coffee causes cancer—at least not as Western medicine now understands the connection. Unlike the deductive thinking of Western science, which looks for absolute truth and linear cause-and-effect relationship between one particular substance and our health, Chinese medicine's inductive reasoning looks for patterns and tendencies within the whole picture. Just as coffee on its own doesn't create climate change, coffee by itself doesn't always cause cancer. However, coffee is clearly stimulating and does create heat internally. Coupled with its dampness, its overstimulation can be hard to clear from the body. Within the context of a country so amped up that's it's overheating the whole planet, this contributes to an overheated internal condition. When this heat has cooked off much our fluids, and latency is lost, conditions that come from heat begin to appear, including cancer.

The Deeper Issues of Cancer

In addition to the physical causes of cancer, there are also deeper and philosophical causes at play. Not only do alcohol, tobacco, coffee, and too much meat have warming effects, heat and excess growth is created within us by how we see the world. Our perspective of continuous activity and seeking newness speaks to our overemphasis on Yang and our lack of Yin. Our desire for continuous economic growth, our valuing of youth, and our view of nature as a place of continuous conflict speaks to being stuck in Wood.

As we've discussed in previous chapters, this very same dynamic of too much Yang, not enough Yin, and an excess of Wood is appearing within us and all around us. It's part of our beliefs about the world and the cultural systems we've created. We're taught that continuous economic growth is both possible and desirable. From the view of modern biology, all individuals and all species are continuously looking for a competitive advantage over everyone and everything else—a warring nature view of the world.

Medically speaking, this Yang excess and overemphasis on Wood can make it seem like fighting and killing what's causing our symptoms is the best, and maybe even the only, valid medical approach. However, our lives and medicine make up more than just one half of the Yin-Yang circle or one of the Five Phases.

Thankfully, there are acupuncture treatments that can clear out heat at the deepest level of who we are. These include treatments at depths that are analogous to the Western understanding of DNA and RNA—in Chinese medicine, these would be at the level of our jing. The modern text *Advanced Acupuncture* by Ann Cecil-Sterman, a long-time student of Jeffrey Yuen, discusses the historical and contemporary uses of the Eight Extraordinary Meridians. As Cecil-Sterman describes, the "Eight Extras" work at a deeper, generational level than the more commonly used pathways of energy that correspond to our internal organs. She writes that a seminal historical text, the *Nan Jing: Classic of Difficulties*, describes the depth of the Eight Extras and how they can't be accessed through the use of the primary channels alone. In particular, she notes the importance of their use in the treatment of sicknesses that affect DNA, including cancer.[11]

Using strong substances to treat serious conditions also has a long history in Chinese medicine. Potent and toxic medicines like Scorpion and Centipede are prescribed for dramatic symptoms, including internal wind, such as seizures, and to forcefully clear out toxins, such as cancer.[12] The use of Centipede in particular dates back several thousand years in written form to the oldest existing Chinese herbal text, the *Sheng Nong Ben Cao Jing (Divine Farmer's Materia Medica)*.[13]

In contrast with treatments like chemotherapy and radiation, acupuncture and herbal medicine can clear out the heat that is a root cause of cancer while also increasing the cooling effects of Yin and promoting overall health. From several thousand years of clinical application, there are well-developed and time-tested ways of using very strong substances and deep-reaching treatments without further compromising people's well-being.

For example, when using substances like Centipede and Scorpion, sweet-tasting licorice is often added to formulas to make them more digestible and significantly reduce toxicity.[14] Spicy fried ginger is also used to help assimilate harsh substances by strengthening the digestive system and preventing it from getting stagnant. This balance—of strongly clearing out heat while mitigating the side effects and promoting well-being—is a medical approach that is centered on peace rather than war. It is especially important when treating a person who has cancer, which is likely due to significant heat and equally significant loss of coolant, to actively promote health.

An additional aspect of this approach includes herbs that also clear heat and increase coolant. Increasing Yin helps to balance an overheated condition, and herbs like processed *Rehmannia* root and *Scrophularia* root clear heat at a deep level and nourish Yin simultaneously.[15] When combined with substances like Scorpion and Centipede and herbs like licorice and ginger, *Rehmannia* and *Scrophularia* are part of potent formulas that clear heat, increase coolant, protect digestion, and strengthen Qi.

Of course, each herbal formula or acupuncture treatment is best customized to each person and their condition. Because each of us are unique individuals, whether we have the flu or cancer, how we developed that condition is personal. The overgrowth and spread of unhealthy cells fits well with the Chinese idea of heat, but understanding how we accumulated that heat, how severe it is, and which organs are involved needs to be diagnosed on a person-by-person basis. Just as it's essential to understand the underlying causes of our warming planet, the most effective acupuncture treatments and herbal formulas address our symptoms as well as their individual root causes.

The need for a customized approach is true not only for treatments but also for lifestyle changes. As discussed, Western medicine often promotes eating well, maintaining one's weight, and exercising to reduce the likelihood of cancer. While certainly relevant, this one-size-fits-all approach has significant limitations as it is based on general trends and not on the condition of the specific individual.

The Limits of More in Treating Cancer

One example of the limitation of general statements is when we focus on the amount of exercise to get. While exercise is undoubtedly important from both an Eastern and Western point of view, more is not necessarily better. Many of us could benefit from the increased movement and deep breathing that exercise provides. But, there are many of us who could benefit from less. As we talked about with Yin and Yang, just because *some* of something is helpful, *more* of it doesn't necessarily promote health.

Exercise promotes the movement of Qi and blood within the body and can directly strengthen muscles, tendons, ligaments, and the respiratory and circulatory systems. How much exercise we need to promote health depends on many variables, however—age, overall health, stress levels, and the amount of Qi we have. In a Yang-excessive culture, something as health-promoting as exercise can be taken too far and instead promote sickness.

When we sweat, we create heat. This is because one way the body tries to get rid of excess warmth is through the skin. In addition to warming us up, the regular loss of fluids through heavy sweating can also dry us out. As with most lifestyle choices, the way and the degree that this affects us depends on our internal condition and the condition of our Yin and Yang in particular.

Since we're already living overstimulated lives, one possible result of too much vigorous exercise is our heat rising and coolant decreasing. From this perspective, it's not a sign of a good workout to get sweaty all the time. It's often more an indication that we're overheating ourselves and cooking off our fluids.

From an Eastern view, fluids are precious substances. Physically, they are essential to keeping things cool and soft, and prevent our muscles, tissues, and organs from becoming hard, getting hot, and drying out. Fluids have a similar effect mentally and emotionally, helping to keep our mind and responses to the world flexible and in balance. When we regularly decrease our fluids through too much vigorous exercise, we're literally sweating out our internal coolant. In other words, *too much vigorous exercise can contribute to, and even create, Yin-deficient heat.* That's right: this is the same Yin-deficient heat that is warming the planet and creating the conditions for cancer.[16]

Having spoken to many people in the treatment room about the possible results of too much exercise, I can imagine what you might be thinking. Exercise is good for us, sweating is a way to clear out the body and detoxify, working out makes you feel good, and so on. From a Chinese viewpoint, all of these things can be true—but just like drinking coffee, the issue is context.

We live in a Yang-excess, Yin-deficient country. Within our hot, dried-out culture, doing activities that create more heat and less Yin is just contributing to more of the same. This includes a rapidly warming planet and a staggering number of cancer diagnoses.

As with all things, we need a balance that is specific to our own internal condition. For most of us, moderate regular exercise—with no sweating or only moderate perspiration—is likely to best promote health. Too much activity, with too much sweating, will not balance Yin-deficient heat.

A Proportionate Response to the Lessons of Cancer

Similar to climate change, it's imperative to have a proportionate response to cancer. Clearly, 125 million possible diagnoses and 72 million possible deaths are a severe problem. These projected numbers can serve as a wake-up call—the seriousness of the condition is one way our bodies and our cells are trying to get our attention.

Even though the diagnosis can create fear and bring up questions about our own mortality, cancer is not the enemy, and waging war is not the only option. Cancer is no more and no less than the proliferation of unhealthy cells, which are the result of too much heat, an excess of unhealthy growth, and not enough Yin.

With our understanding of Chinese medicine, we can recognize that the individual and internal issues of cancer are a direct reflection of the external global crisis of climate change. In other words, the heating and loss of Yin that create cancer mirrors the increase of emissions of greenhouse gases and the decreasing ability of the planet to sequester them. The overgrowth of cancer cells from an excess of Wood and the increasingly severe storms of climate change mirror our quest for continuous growth.

How we approach the significant diagnosis of cancer speaks to how we look at the life-threatening condition of climate change. Just as waging war on climate change only leads us farther down the path we're currently on, fighting cancer continues to promote an excess of Wood in our country. Certainly, there can be times when the overgrowth of unhealthy cells has progressed to the point where radiation, chemotherapy, and surgery may be viable options. Waging war, however, doesn't lend itself to listening to what our cells are trying to tell us. If we simply try to get rid of the symptoms rather than understanding their deeper causes, it's often much harder to hear the messages they're trying to tell us.

The overgrowth of cells is trying to tell us that we're so hot and overstimulated that 125 million of us may be diagnosed with cancer and 72 million may die from the diagnosis if our lives, beliefs, and cultural systems continue in the direction they're headed. Our severe overstimulation is a reflection of the advanced heat we've created in the climate, which may be at a tipping point of dramatic, rapid warming.

While it might be hard to hear, part of the issue with the extraordinary number of cancer diagnoses is that cancer is a big business. A National Institutes of Health study indicates that in 2010, the cost of cancer treatments in the United States was just under $125 billion. *That's $125,000,000,000 spent on Western treatments for one condition in one country in one year.* The study also estimates that this cost will increase to at least $158 billion by 2020, an increase of twenty-seven percent in ten years. The research also indicates that, based on current trends in cancer treatments, the cost could reach $207 billion yearly, a sixty-six percent increase in a decade.[17]

What we're seeing with the dramatic number of cancer diagnosis and the equally dramatic costs of Western treatments is the intersection of three related views of the world. *First,* we have a view of separation. This is apparent in medicine, as organs are viewed as isolated from each other and the physical, mental, and emotional parts of our lives are also viewed as distinct. Additionally, this perspective holds that what we see happen in nature is separate from what happens to us. As a result of this perspective of isolation, it's difficult, if not impossible, to see the patterns that create the underlying causes of cancer. As is common with Western medical treatment with sickness of all kinds, we're waiting for the problem to occur before we treat it.

Second, we're encouraged to be comfortable with conflict and waging war. Western science promotes a view of nature as a place of continuous struggle and self-interest, and suggests that waging war on cancer is the only viable approach. Not only does the emphasis on killing cells and cutting out body parts affected by cancer discourage us from listening to our symptoms, the Western approach can also create the diagnosis it's intended to treat.

From working with people with a cancer diagnosis, we've seen at our clinic the effects of the Western treatment. While they're obviously intended to get rid of the cancer, chemotherapy and radiation also come with a long list of other effects. Though they are referred to a side effects in Western medicine, they often do not happen on the side at all but affect people front and center. I've had patients explain to me, and have seen from Chinese diagnostics, how the wide range of symptoms associated with radiation and chemotherapy is affecting them. They include nausea and vomiting, loss of appetite, second- and third-degree burns, joint and muscle pain, sleep disturbances and insomnia, loss of energy, and even a loss of the will to live. The American Cancer Society states that the likelihood of developing secondary cancer varies with higher drug doses, longer treatment time, lifestyle choices, and other patient-specific issues. However, the synopsis of the research states clearly that both chemotherapy and radiation are associated with higher rates of cancerous tumors and several types of leukemia in particular. Specifically, citing a paper from the National Institutes of Health,

> Radiation therapy was recognized as a potential cause of cancer many years ago. In fact, much of what we know about the possible health effects of radiation therapy has come from studying survivors of atomic bomb blasts in Japan. We also have learned from workers in certain jobs that included radiation exposure, and patients treated with radiation therapy for cancer and other diseases.[18]

There is also an established link between cancer and chemotherapy. The American Cancer Society summarizes, "Chemo is known to be a higher risk factor than radiation therapy in causing leukemia.... Some solid tumor cancers have also been linked to chemo treatment for certain cancers, such as testicular cancer."[19] Waging war on the growth of cells can create collateral damage—not only on the digestive system, skin, and overall energy, but also on the very diagnosis it's intended to treat.

Third, we have an often unquestioning belief in continuous growth. With our overemphasis on Yang and Wood, we're encouraged to gauge the well-being and health of our economic system based on its expansion and growth. This measure of economic health is also applied to the individual companies and institutions that function within the larger system, including those that provide medical services. Coupled with the encouragement to focus on economic self-interest, we've created a structure in which it's good for business when people receive a cancer diagnosis. It's clear from the hundreds of billions of dollars being spent that the current economic incentive for Western medicine is not to reduce the number of cancer diagnoses. In fact, the more people diagnosed with cancer, the more money there is to make. As we've seen above, the already incredible amounts spent are very likely to increase—rather than encouraging health, Western medicine is being paid for sickness.

To clarify, I'm not implying that Western medical institutions and practitioners are not trying to help. We treat many Western practitioners and administrators at our clinic and are friends and colleagues with many others. I have no doubt that many of them are genuinely committed to helping patients to the very best of their ability. However, the cultural systems we've created, and the assumptions on which they're based, are contributing to the millions of cancer diagnoses we're seeing and the hundreds of billions of dollars we're spending to treat the condition.

So what's the alternative? From the view of Chinese medicine, it is to prevent the disease from occurring in the first place and to actively promote health individually and culturally. In addition to the economic and medical issues that arise from waiting to treat cancer until it has already manifested, how we live, what we believe, and what we value are warming our internal environment as well. As things heat up internally, our ability to hold symptoms in latency is lost, and unhealthy cells can begin to spread within the body.

We are a mirror of nature and it is a mirror of us. Just as global warming is trying to get our attention, so is the growth of our cells.

The Tale of Two Diagnoses: Transformation

We're free to choose how we respond to climate change. It's also up to us how we react to a cancer diagnosis. Once we have the information about the growth of unhealthy cells or about our warming planet, what we do with it depends on how we see our lives and view the world. Data about the growth and movement of cells or about the rates of increase in greenhouse gases is only part of the discussion—the other essential part is how we respond.

To help us examine the possible range of responses, let's look at two patients I treated at our clinic. The first is a middle-aged man whom I'll call Ben. Ben, who lives in a small rural community, is tall, strong, and stout from a life of hard physical work, which included long hours outside year-round as a farmer.

He came to see us after his Western doctor ordered a series of blood tests based on his symptoms. He was having regular night sweats, had moderate-size lumps on his neck, and had been tired for several years. Because he'd had these symptoms for some time, he'd also had similar blood work done three years earlier. Then, his white blood cell count was 25,000 per microliter. When he came to see us five years later, it was 48,000.[20] As Ben told me during our appointment, his Western doctor told him that a fifty percent increase in white blood cells is one indication of chronic lymphocytic leukemia (CLL).

According to the National Institutes of Health, CLL is a type of cancer of the white blood cells, called lymphocytes. These cells are found in the bone marrow and other parts of the body. CLL causes a slow increase in a certain type of white blood cells called B lymphocytes, or B cells, which can spread through the blood and bone marrow. CLL can also affect the lymph nodes or other organs such as the Liver and Spleen. In addition to elevated blood count, Ben's other symptoms of sweating, enlarged lymph nodes, and continuing fatigue are also symptoms of CLL.[21]

Ben also told me that since his numbers hadn't quite increased by the requisite fifty percent to warrant Western treatments, he was told to come back in a few months to have his numbers checked. Once they reached 50,000, his doctor said he could start with chemotherapy.[22] Ben had seen several members of his close-knit community go through similar treatments for the same condition, and saw first-hand the effects that come with chemo that we discussed earlier. He came to us looking for other options.

CLL fits well with the Chinese understanding of heat. Ben's night sweats were coming from Yin-deficient heat, as he didn't have enough coolant to maintain his temperature later in the day. The lumps on his neck were where the heat was concentrating, and the overgrowth of white blood cells was from overstimulation at a cellular level. In particular, blood cancer corresponds very closely to the Eastern diagnosis of heat in the blood. At this stage of progression, the inflammation had reached a deep level. From the tradition in Chinese medicine that specializes in heat, the *Wen Bing,* this blood heat is a final stage of the progression of sickness.[23]

As we do with almost all patients, whether they have a relatively mild condition or life-threatening diagnosis, I talked to Ben about his lifestyle. I told him that if he really wanted to give Chinese medicine the opportunity to help, he had to participate in the process. Specifically, he needed to change the parts of his life that were creating heat and cooking off Yin. Being the hard-working man that he was, that meant doing less and resting more.

He agreed to hire more help and reduce his usual sixty- to seventy-hour work week to thirty-five to forty hours. He also agreed to allow his employees to take on more of the demanding work at the farm so he could concentrate on jobs that were less taxing physically. He also agreed to relax more and do less, and increase his sleep from six or seven hours to nine or ten per night. He agreed to eliminate foods that create heat, which for him meant eliminating chicken, while maintaining his strength by eating more pork. He also cut out all processed sugar, as it's warming and stimulating, and increased foods that were cooling, including a wide variety of leafy green vegetables. In place of the several sodas he was drinking each day, he substituted water and green tea, both of which are cooling.

For the first month, he received acupuncture twice a week, and for the second month, he received treatments weekly. During these eight weeks, he also took a moderately high dose of a Chinese herb formula three times a day to treat heat in the blood, with additions to clear out toxins. The formula included the herbs we discussed earlier that clear out heat and increase coolant—*Scrophularia* and prepared *Rehmannia*—as well as ones to clear out toxicity—Scorpion and Centipede.[24]

After two months of diligent lifestyle changes and consistent acupuncture and herbs, Ben was doing much better. His night sweats were gone completely and the lumps on his neck had decreased significantly. He described how his energy had returned and how the mental fog he had before starting treatment was also greatly improved. He had also lost thirty pounds, returning him to the trim weight he had been in high school. As a sign of the improvements, he told me that he hadn't felt this good in twenty-five years. His pulse and tongue told a similar story—his heat had decreased significantly and Yin was increasing.

As he felt so much better, he went back for another set of blood tests. Confirming how he felt as well as his Chinese diagnosis, the test results showed very significant positive changes. In eight weeks, his white blood count had gone from 48,000 to below 12,000. This was over a seventy-five percent decrease from his most recent result two months earlier and a fifty percent decrease from his numbers three years earlier, taking him well out of the range of a leukemia diagnosis. According to Ben, his Western physician was so surprised by the results that he ordered another set of blood work, assuming that there must have been a mistake with the testing process. This time, the results after another week of acupuncture and herbs showed a drop of another 300 points.

In addition to the obvious health benefits from his treatment, there were significant economic benefits as well. The total of twelve acupuncture treatments and eight weeks of herbs cost about $1,240. While there are many different forms of chemotherapy to treat CLL with varying costs, a study in *The Oncologist* states the wholesale price for a twelve- to twenty-six-week treatment cycle of a newer medication is between $35,000 and $120,000. The study also indicates that these amounts don't include the cost of administering the drug or the price of other medications associated with chemo.[25]

Additionally, unlike chemotherapy, which often comes with a long list of difficult negative reactions, Ben's thinking cleared, his night sweats went away, the lumps on his neck decreased, he lost thirty pounds, and he felt better than he had in years. Not only did Chinese medicine address Ben's heat, which had created the overgrowth of unhealthy cells, its ability to wage peace had many other benefits as well.

The Tale of Two Diagnoses: Management

The second patient I want to tell you about is Molly, who was in her mid-forties when she came to see us after being diagnosed with breast cancer at about the same time I was working with Ben. She was worried and somewhat scared as we talked about the Western prognosis and how Chinese medicine could be of help.

When I asked Molly what kind of help she was looking for, she responded, "I just want my life back." Coming from a Western viewpoint, this is understandable. When confronted with something like cancer, one response is to try to create a sense of stability, where we continue with the things we've been doing at home, at work, or at school. But from an Eastern view, how we are living often contributes to our diagnosis. In asking Molly about her life and what she wanted to go back to doing, she told me that she often worked long, stressful hours at a hospital, exercised vigorously most days, and was generally a busy person. She also enjoyed drinking wine and coffee each day.

She had discussed the possibility of surgery with her Western doctor but was planning on having radiation and possibly chemotherapy. While I said that Chinese medicine would likely be of significant help with the effects of both chemo and radiation and could assist her before and after surgery, I also wanted to discuss some of the root causes of where the cancer was coming from.

I explained heat, a decrease in Yin, and a loss of latency, encouraging Molly to do less and rest more, as well as to get enough sleep. I also spoke about how it was important with a cancer diagnosis to protect the fluids and not sweat them out. We discussed the vital importance of diet, and the Chinese medicine understanding of how certain foods and drinks were cooling and others warming. In particular, I talked about the stimulating nature of coffee and alcohol and how they're connected to the overstimulation of unhealthy cells.

After about fifteen minutes, she looked at me again and said, in slightly different words, "I want my life back." I told her that I understood and agreed to help her as much as I could as she went through her Western treatments. In hindsight, it's not surprising that Molly only came in for a few treatments and did not agree to take herbs. As she made clear, she was looking to keep things the way they were and get rid of her symptoms. Of course, it's her right as a patient to decide what treatments she receives and what changes she makes. She was approaching her health care and her cancer diagnosis from the perspective common in the United States and the West—get rid of the problem by killing the cells and possibly cutting out the affected area, while continuing what we're doing.

In looking at the experiences of Ben and Molly, Ben was committed to changing many parts of his life, including the hours he worked, the amount he slept, what he ate and drank, and how he approached his life. Molly chose to try to maintain her hours at work, her diet, her exercise, and her lifestyle.

While it's not a value judgment about how they approached their diagnosis, there are clearly consequences to the different responses. From a Chinese medicine perspective, Ben made changes to his life, received acupuncture treatments, and took herbs to treat the root cause of his diagnosis. For Molly, the approach was to continue to create heat, cook off Yin, create toxicity, and then kill off and cut out the results of these root causes.

We have a similar choice with climate change. How we view the increases in emissions, methane bubbling from the ocean floor, and melting permafrost speaks to how we see the crisis. The usual Western discussion of symptoms—from cancer to climate change—involves how to get rid of the problem. With cancer, it's how we kill the cells that are causing the problem as we get back to our lives. With climate change, it's how we reduce emissions and sequester more carbon while keeping in place our cultural systems and the beliefs on which they're based. But just as a significant diagnosis like cancer comes from the condition of our internal environment, climate change comes from deeper issues within us and within our country and culture.

If we engage with the deeper issues of global warming, not only is there the possibility for a stabilization of the climate, but there is also an opportunity for our own healing as well. Just as Ben's temperature regulated, his thinking cleared, his energy returned, and he lost weight, we could see long-lasting changes to our well-being in our own lives and in our country and culture.

Along with our rapidly warming planet and the extraordinary rates of cancer, there's another, person-specific way to begin to understand the state of our Yin and Yang. As we'll talk about next, tongue diagnosis is an accessible method to see concretely the condition of our internal warmth and coolant, and the state of our Five Phases and internal organs.

CHAPTER 10

Internal Climate Change

Tongue Diagnosis

Rather than just taking my word for it, let's talk about how you can look at the condition of your own Yin and Yang and your Five Phases. A primary method used in Chinese medicine to evaluate the state of our internal heat and coolant and the condition of our internal organs is tongue diagnosis. If we're to address the underlying causes of climate change, it's essential that we understand that what is occurring with the climate is occurring within us as well. The degree that the global environment is becoming destabilized is directly proportional to the destabilization of our own internal ecology.

We know there is heat from increasing greenhouse gas emissions and the release of methane from bogs and the ocean floor. We also know there is a loss of global coolant from deforestation, melting ice sheets, and the acidification of the ocean. As a result, we've seen storms like the ones we talked about in chapter 1, which turned the river outside our clinic from a quiet murmur into a raging Class V rapid overnight. These dramatic and rapid changes in the environment around us mirror an equally dramatic and important change to our internal environment.

In essence, the small picture is a mirror of the big picture—as a small part of nature, we are a reflection of the whole. Likewise, because our tongue is physically a small part of our bodies, its condition is a mirror of the bigger picture of what's happening within us—physically, mentally, emotionally, and spiritually. This time-tested holographic understanding of the world and our relationship to it allows us to see the patterns that affect our health and the health of the planet. Tongue diagnosis is an easily accessible, effective, and deep-reaching method to understand our own internal condition and what is, and isn't, promoting health.

Tongue diagnosis can be understood as a map of our internal terrain. We don't have to rely exclusively on MRIs or x-rays to know the condition of our internal organs. Their state manifests externally, in great detail, on our tongue. The size, shape, color, coating, and texture of our tongue tell us what's happening within us. Depending on what the tongue looks like, we can determine whether we're hot or cold, damp or dry, or have internal wind. The tongue also tells us the condition of the Qi throughout our body. Additionally, in Chinese medicine, different qualities observed on different parts of the tongue indicate which organs are being affected and how.

The forward to *Atlas of Chinese Tongue Diagnosis,* an in-depth modern text, states that the first person thought to have used tongue diagnosis is Bian Que, a renowned Chinese physician from the Warring States era (471–221 BCE). The atlas also attributes the first written reference to tongue diagnosis to the seminal Five Phases text we discussed earlier, the *Nei Jing.* Additionally, the first text to deal exclusively with tongue diagnosis dates to 1341 CE and was published by Du Qing-Bo, based partially on the work of another author.[1]

Before we get into the specifics of tongue diagnosis, let's begin with the well-developed logic behind it. In Chinese medicine, when we say "the body," we are referring to the torso. While the arms, legs, and head are obviously important, they are considered less essential as they don't house any organs thought to be of primary importance.

That's right—the brain, which is given such prominent attention in Western culture and medicine, is not considered a primary organ in Chinese medicine. It's categorized as a secondary organ—a curious organ that doesn't function the same way as other organs that are directly connected to the Five Phases. The rationality and thinking associated with the brain is recognized as important in Chinese medicine, but it isn't as essential as the actions of the primary organs.[2]

With that said, the tongue is a small, external map of our larger, internal terrain—particularly the condition of the body, or the torso, which houses the primary organs. The front of the tongue is associated with our upper torso and our chest, the middle of the tongue with our middle abdomen, and the back of the tongue with our lower abdomen. Also, a tongue showing internal health is pink, flat, and smooth, without curled edges, cracks, bumps, or dips. It has a consistent, oval shape and a thin, clear coating, without any yellow or white on the surface. Not surprisingly, with the long history of Chinese medicine, there is some difference in interpretation of tongue diagnosis. For clarity's sake, we will discuss the perspective I use clinically, which includes a small change from what is presented in the *Atlas of Chinese Tongue Diagnosis.*

Understanding Our Inner Ecology

Associated with the chest, the very tip of the tongue specifically corresponds to the Heart, the Pericardium, and the Lung. The center and middle of the tongue is associated with the digestive system, specifically the Stomach and the Spleen, as they are physically located in the middle of the abdomen. The middle sides of the tongue relate to the Liver and Gallbladder, as they are located on the sides of the abdomen. The very back center of the tongue corresponds to the Kidney and Bladder, and the sides of the back of the tongue are associated with the Small Intestine and Large Intestine, as they are located in the lower torso.[3]

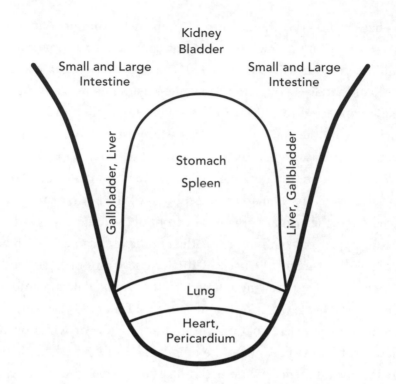

Now that we've established the map of the tongue, in which certain parts are associated with their respective primary organs, let's discuss the various qualities that can be found on the tongue. Because tongue diagnosis is an in-depth medical evaluation that requires years of training to apply thoroughly, we won't be able to cover everything that you might encounter during tongue diagnosis. Let's focus on some of the diagnoses we see regularly at our clinic and that are related to the condition of our country and our climate.

Color: When the organs and Qi are in balance, the whole tongue has a light-pink color, from front to back and side to side.

- *Redness* on the tongue indicates excess heat. A severely red tongue that is dark or crimson red indicates more advanced heat.

- *Raised red dots* indicate advanced heat as well as toxicity. This is a sign of excess warmth pushing itself off the surface of the tongue—the dots show that the heat has accumulated to toxic levels. When the dots are very prominent, and they look like small growths rather than dots, this indicates an even more significant accumulation of heat and toxicity, called *fire toxins.*

Coating: When there is a healthy balance of moisture internally, the tongue has a slight, clear coating.

- A *dry tongue* occurs when there is a lack of fluids and a lack of Yin.

- When there is internal heat, a *thin, smooth, white coat* will appear, indicating that the fluids that were once clear are being cooked. This coating comes from what Chinese medicine calls dampness, which is an unhealthy accumulation of too much fluid.

- The next stage in the progression of heat and dampness is a *smooth yellow coating,* which indicates more heat than the white coating. The amount of yellow indicates the degree of heat, and the thickness of the coating indicates the amount of dampness.

Texture: When the organs are in balance, the tongue is smooth and flat throughout its entire length and width.

- *Vertical cracking* on the tongue indicates dryness, just like exposed soil when there's too much sun without enough rain.

- *Horizontal cracking* indicates Qi deficiency. Part of what our energy does is hold things together. When there's not enough Qi, things begin to pull apart and separate, creating horizontal cracks.[4]

Size/Shape: A tongue indicating internal health is evenly rounded at the front and has a consistent width and thickness.

- A *pointy* tongue that is elongated at the tip indicates excess heat. It very often accompanies a red tip—both indicate heat in the Heart.

- *Scallops* on the side of the tongue indicate a lack of Qi. Scallops are wavy indentations that occur where the tongue is pressing against the side of the teeth. Because our energy is responsible for holding things in place, when it's deficient, the sides of the tongue get puffy and push against the teeth, creating scallops.

- A *puffy tongue* also indicates Qi deficiency. Just as the sides of the tongue get scalloped when there's not enough Qi to hold it in place, a puffy tongue can also come from a lack of energy.[5]

- A *dip* in the tongue indicates Yang deficiency, as there's not enough deep strength to hold the tongue up. A dip in the back indicates jing deficiency, as this part of the tongue is associated with the Kidney, which houses the jing.

- A *short tongue* also indicates jing deficiency. This occurs when there is a complete absence of the back of tongue. As it's associated with the Kidney, when the back of the tongue is missing, it implies a significant lack of jing.

- A tongue that's turned to the side is called a *deviated tongue,* which indicates internal wind. This occurs because there's an excess of movement and a lack of stability in the Liver and Gallbladder, causing the tongue to move to either side. The degree of deviation speaks to the amount of internal wind.

Evaluating Our Inner Ecology

To look at the state of your internal ecology by examining your tongue, find a place with a well-lit mirror. Because there's a great deal of important information that you can derive from a tongue diagnosis, I suggest having a pen and a piece of paper. To start, simply stick out your tongue as far as you can without creating tension in your mouth.

It usually takes looking at your tongue several times to understand what you're seeing. It is important only to look at your tongue for fifteen to twenty seconds at a time. If you look longer than that, it will start to change shape and color from the effort. To begin, first look at the tip of your tongue and note its color, shape, coating, and texture. Write down what you see.

As the tip of the tongue is associated with the Heart and Pericardium, which are part of the Fire phase, it's prone to getting hot. Just as the hottest place in a room is often near the ceiling because heat rises, the hottest place in the body is often the chest, and the Heart in particular.

Tip of the Tongue

- If the tip of the tongue is *red,* this indicates heat in the Heart and Pericardium.

- If the tip of the tongue is *pointed,* this also indicates heat in the Heart.

- If the tip of the tongue has raised *dots,* this indicates toxicity and more advanced heat in the Heart and Pericardium.

- If the tip of the tongue is *dry, without a coating,* this indicates a lack of fluids and Yin deficiency in both organs.

- If the very tip of the tongue has a *vertical crack,* this also indicates Heart and Pericardium Yin deficiency. The depth and width of the crack indicates the amount of dryness.

One of the most common presentations we see at our clinic is a red tip of the tongue with raised dots. In the undergraduate Chinese medicine classes I teach, I often have students look at the tongues of others on campus as part of their assignments. After a few weeks of looking at a few dozen tongues, they often come back to class and say that almost everyone has a red tongue tip and ask if this is "normal."

As we've discussed before, just because something is common doesn't mean that it's a sign of health. In our country, which is so overheated that it's destabilizing the planet, many things that are considered usual are often pathological. Just because the tip of your tongue is red, as are the tongues of nearly everyone around you, this doesn't mean a red-tipped tongue is the "new normal." Rather, it is yet another indication of how much heat we each have individually and how pervasive it is in our culture.

When we have heat in the Heart and the Pericardium, this means our Fire phase is overstimulated. In particular, it indicates that our expression of who we are, which occurs through the Heart, is lacking internal peace. Heat in the Heart can also contribute to a wide variety of other symptoms, including insomnia (particularly the inability to fall asleep), anxiety, chest pain, heart palpitations, and other circulatory conditions. Given that the state of the climate is mirrored in our internal conditions, it's not surprising that we see so many people with these symptoms at our clinic.

Having a red tip of the tongue, especially when coupled with a crack that indicates a lack of Yin, can also mean that your communication with others has been overheated as well. As we talked about in chapter 8, on Fire, the amount of emails, texts, and tweets many of us send and receive each day speaks to how overstimulated we've become. And the use of the devices themselves also contributes to the buzz of a lack of coolant and too much warmth.

Behind the Tip of the Tongue

Next, look at the front of the tongue, which is the area immediately behind the tip. This is the area associated with the Lung, which is physically located in the chest, next to the Heart.

- If the *front of the tongue is red,* this indicates heat in the Lung.

- If the *front of the tongue has raised red dots,* this shows advanced heat and toxicity in the Lung.

- If the *front of the tongue is dry and lacks a coating,* this indicates a lack of fluids and Yin deficiency in the Lung.

- If the *front of the tongue has a vertical crack,* this also indicates Yin deficiency in the Lung, a more advanced stage of dryness than a lack of coating. As with the Heart and all of the other organs, the depth and width of the crack indicates the amount of dryness.

As heat rises upward, and as the Lung is in the upper torso along with the Heart, it's also common for the front of the tongue to show signs of heat and dryness. This excess warmth can start in the chest or originate in other organs and rise up into the Lung and Heart. When there's heat in the Lung, the organ itself is overstimulated and possibly inflamed. This can correlate to a wide range of respiratory issues, including minor ones like chronic cough, mild shortness of breath, and chest tightness. It can also include significant diagnoses, such as certain types of asthma, emphysema, chronic obstructive pulmonary disease, and other long-term respiratory conditions.

When the Lungs are hot and Yin deficient, this can also mean that the connection to the spiritual parts of our lives is overstimulated and dried out. As we talked about in chapter 7, one of our fundamental needs to live a healthy life is a direct experience of something greater than our everyday lives. When the Lungs are hot, it's more difficult to experience a lasting sense of inspiration. Heat often cooks off the fluids, and this Yin deficiency in the Lung makes it harder to absorb the Qi from heaven that we inhale with each breath.

The Middle and Center of the Tongue

Now look to the center, middle of your tongue. This area corresponds to the condition of the middle of our abdomen—our digestive system, in particular. In Chinese medicine, this includes our Stomach and Spleen. Similar to their functions as described by Western medicine, the Stomach is responsible for taking in food nourishment as well as for the majority of digestion. From an Eastern view, it also helps distribute the Qi we garner from eating food to our other organs. The Spleen works with the Stomach to distribute this food energy, and it also creates blood and lifts Qi upward throughout the body. The Spleen is also responsible for the strength of our muscles and contributes significantly to our overall mental and physical vitality. The Spleen is the organ most closely associated with our thoughts and our thinking process in general. Together, the Stomach and Spleen turn the food we eat into a major source of our day-to-day energy and strength.[6]

When looking at the middle and center of the tongue, it's possible that you'll see similar things to what appears on the front and tip.

- If the *middle, center of the tongue is red,* this indicates heat. The redder this area is, the hotter things are internally. If there are *raised red dots,* this indicates more advanced heat as well as toxicity.

- If the *middle, center of the tongue is dry and without a coating,* this shows dryness in either organ.

- If *the middle, center of the tongue is cracked vertically,* this indicates Yin deficiency in the Stomach and Spleen.

- If the *middle, center of the tongue has a smooth white coat,* this indicates dampness from heat. If there's *a smooth yellow coat,* this indicates more advanced heat. The darker the color of the tongue's coat, the more intense the heat. Likewise, the thicker the coat, the greater the dampness.

It's also possible that you'll see characteristics less common on the front and tip.

- If the *sides of the middle of the tongue are scalloped,* this indicates Stomach and Spleen Qi deficiency. As we'll discuss further below, the middle sides of the tongue are associated with the Liver and Gallbladder. However, when the sides are scalloped, it's due to the middle of the tongue pushing outward, causing the sides to push against the teeth, creating scallops.

- If the *middle, center of the tongue is indented,* this indicates Spleen Yang deficiency. The Yang helps to hold things up, and when we're Yang deficient, things start to drop. This is a deeper level of tired than mere Qi deficiency.

Just as the tip and the front of the tongue can be red and cracked, so can the middle of the tongue. Heat and dryness in the Stomach and Spleen are related to a long list of physical symptoms, including indigestion, acid reflux, stomach ulcers, Crohn's disease, irritable bowel syndrome (IBS), too much or too little appetite, and a wide variety of digestive issues. Too much heat in the Stomach and Spleen can also contribute to sense of being overstimulated physically. As it is a major part of our day-to-day energy, heat in the digestive system often leaves us feeling amped up.

Mentally and emotionally, a hot and dry Stomach or Spleen overstimulates our thoughts. As the Spleen is associated with thinking, too much heat and too little Yin often mean an overly busy mind. In our overstimulated culture, it's not surprising that so many of us have a hard time controlling our thoughts.

At our clinic, many patients describe how hard it is to turn their mind off at night and fall asleep. A major contributor to this type of insomnia is too much heat and not enough Yin, especially in the Spleen. If you tried the exercise suggested in chapter 7, on Metal, and had a hard time sitting still and relaxing your mind, Yin-deficient heat in the Spleen and Stomach is one possible cause.

With all the overstimulation and lack of Yin in the world around us, it also makes sense that so many of us are tired as well. Many people we see at our clinic have scallops on the sides of the tongue as well as a dip in the middle. This Qi and Yang deficiency sometimes corresponds to heat in the Stomach and Spleen as well as redness throughout the tongue. This dynamic of overstimulation and tiredness is quite a common one in our era of climate change. Just as the planet is reaching its limits of what it can sustain, our Qi is similarly reaching its limits. When we're tired, what keeps us going is our internal heat. As noted earlier, many of us use coffee to keep going when we don't have the energy to maintain our overstimulated lives. But heat is not Qi, and using stimulants to try to drum up energy often leads to consequences that are plain to see on the condition of our tongue.

The Outside Middle of the Tongue

Next, look at the outsides of the middle of the tongue. Physically located in the mid-abdomen, the Liver and Gallbladder also correspond to the middle of the tongue.

- If the *mid-sides of the tongue are red,* this indicates heat in the Liver and Gallbladder. As with the other organs, if there are *red raised dots,* this again indicates increased heat as well as toxicity.

- If the *mid-sides of the tongue are curled,* this indicates Liver Qi stagnation. The Liver in particular is responsible for the smooth flow of Qi,[7] and when its energy is not flowing smoothly, this creates a curling in the sides of the tongue.

- If the *tongue goes to one side,* this is a clear indication of internal wind. The amount that the tongue deviates indicates the amount of wind internally. As it is associated with the Wood phase, internal wind is also associated with both organs connected to this phase, especially the Liver.

Heat in the Liver and Gallbladder, Liver Qi stagnation, and internal wind are very common diagnoses at our clinic. Considering how much we overemphasize the Wood in our country, this does not come as a surprise. When the Liver and Gallbladder are overstimulated with heat and wind, a wide range of physical symptoms results. As we discussed with the patients in chapter 6, on Wood, it can create seizures of all kinds as well as migraines. It can also create a wide range of other kinds of headaches, especially those at the temples and top of the head. Excess heat and wind in the Liver and Gallbladder are also responsible for high blood pressure and for some kinds of dizziness and instability. Internal wind is also an underlying cause of many types of strokes. In one lineage of Chinese medicine, neurological conditions come from an underlying issue of internal wind.[8]

Internal wind can create a feeling of being uprooted and ungrounded, and this is a common dynamic when we're diagnosed with an excess of rising Qi. Too much Liver and Gallbladder heat and Liver Qi stagnation can also contribute to excess anger and rage, as well as persistent frustration. The stories we read about road rage and other acts of sudden violence likely involve significant amounts of heat, a lack of Yin, and stagnation in the Wood. As they're associated with anger, when the Liver and Gallbladder are very overheated and lack adequate coolant, and when the Qi becomes stuck in both organs, there only needs to be one small incident—like someone cutting you off in traffic—to have it explode into rage and violence.

An excess of heat and wind in the Wood can also make it more difficult to have a clear vision for our life. When the Liver and Gallbladder are continuously overstimulated, it's harder to see clearly who we are and what we're here to do.

The Back of the Tongue

Lastly, look at the back of your tongue. As it can be hard to see, make sure to stick your tongue all the way out and that there is ample light. The middle back of the tongue corresponds to the organs that are partially or completely located in lower part of the torso: the Kidney, Bladder, Large Intestine, and Small Intestine. The condition of the middle back of the tongue tells us the condition of the Bladder and Kidney, and the sides of the back of the tongue speak to the state of both Intestines.

Middle Back of the Tongue

- If the *middle back is red,* this indicates heat in the Bladder and Kidney, and *raised red dots* indicate advanced heat and toxicity.

- If the *raised areas are very pronounced and look more like growths than dots,* this indicates significant heat and toxicity—fire toxins—in both organs.

- If it is *dry and without a coating,* this indicates dryness in either organ.

- If *the middle back of the tongue is cracked vertically,* this indicates Yin deficiency in either organ.

- If there is a *smooth white or yellow coat,* this indicates dampness from heat. The thickness of the coat indicates the amount of dampness, and a yellow coat indicates more heat than white.

- If there's a *dip or indentation in the back of the tongue,* this indicates jing deficiency. Unlike other organs, where a dip indicates Yang deficiency, because of the foundational nature of the Kidney, a dip in the back of the tongue shows a lack of the deep reserves: the jing.

- If the *tongue is short,* this also indicates jing deficiency. Rather than a dip, a short tongue means an absence of what would normally be the back of the tongue. This indicates a significant depletion in the Kidney.

So many of us have such excess heat that in addition to redness being common on the front and middle of the tongue, redness and Yin deficiency on the back of the tongue are common as well. Too much heat and a lack of moisture in the Kidney and Bladder mean that our foundational energy is overstimulated. As a result, we feel compelled to keep doing things even though our Qi is no longer able to support our activity levels. If we have more heat than can be cleared from the body, it can accumulate in the Kidney and the Bladder, as they are responsible for cooling thing down and clearing things out.

Heat, dryness, and dampness in the Kidney and Bladder contribute to a number of symptoms in these organs, including Kidney and Bladder infections, Kidney and Bladder stones, and urinary tract infections. Just as the signs of heat on the tongue, including the back, have become common, depleting our reserves has, unfortunately, also become common. Given the way we now live and the amount of fossil fuels we burn, jing deficiency is apparent on the tongues of many patients we treat. Not only is our lack of wisdom prevalent in our cultural systems and the assumptions on which they are based, for many of us, our tongue is telling the same story of jing deficiency.

When jing is deficient, both the Kidney and Bladder are weakened significantly, and as a result, the rest of the organs are weakened as well. The Kidney is responsible for the strength and integrity of our lower back, legs, knees, feet, and ankles; thus, jing deficiency can be a cause of many related physical symptoms in these areas. Lower-back pain and weakness, including vertebral issues, can be related to jing deficiency.

Chronic knee, leg, ankle, and foot pain or weakness can also result from a deep depletion of the Kidney and Bladder. In addition, the Kidney is also responsible for the condition of our bones, so when it is depleted, it can lead to weakness and brittleness of our bones, including osteoarthritis. Jing deficiency is also associated with symptoms of premature aging, including the loss of and graying of head hair, memory and cognition symptoms, and a general decline of vitality.[9]

Aside from these physical ailments, jing deficiency can also make it much harder for us to know who we are and what we're here to do. Without enough jing, it's more difficult to answer the bigger and deeper questions of our lives. Unfortunately, many of us experience the effects of jing deficiency and struggle with being able to sense our unique purpose and how we can contribute to the world.

Sides of the Back of the Tongue

The sides of the back of the tongue tell us about the state of our Intestines.

- If the *sides of the back of the tongue are red,* this indicates heat in the Small and Large Intestines, and *raised red dots* indicate advanced heat as well as toxicity.

- If the *raised areas are more pronounced and look more like growths than dots,* this is a more advanced condition, namely fire toxins.

- If there is a *white or yellow smooth coat on* the *sides of the back of the tongue,* this demonstrates the amount of dampness from heat. A white coat indicates less heat than yellow, and the thickness of the coating shows the amount of dampness.

Considering the way we're living and the amount of impurities we're exposed to on all levels, it's unsurprising that many of us have heat in our Intestines. These impurities include the often discussed substances like pesticides, herbicides, insecticides, chemical preservatives, and artificial coloring and flavors. It also includes all the food that has been genetically modified, sometimes with the genes of other species added.

But the heat in our Intestines is from other sources as well, including the vast amount of information and overstimulation so many of us are exposed to each day. The sheer volume of electronic communication we send and receive daily can tax the capacity of the Intestines to expel the things we don't need.

It's the responsibility of the Intestines to clear things out—physically, mentally, emotionally, and spiritually. When we're exposed to more impurities than the Intestines can handle, significant amounts of fire and toxicity can accumulate, as indicated by large bumps on the back and sides of the tongue.

When there's heat and toxicity in the Intestines, this can contribute to a wide range of intestinal and digestive issues, including Crohn's disease, irritable bowel syndrome (IBS), intestinal polyps and growths, chronic and acute diarrhea, intestinal pain and cramping, and intestinal inflammation of all kinds.

Heat, dampness, and toxicity in the Intestines likely means that accumulation is occurring mentally and emotionally as well. Perhaps we are unable to let go of difficult experiences, or maybe it is hard for us to let go of past or current relationships. Or we may have a hard time letting go of thoughts, beliefs, or patterns that don't promote health. In addition to the physical symptoms that result from accumulation in the Intestines, there are important mental and emotional symptoms as well.

The Opportunity of Tongue Diagnosis

If you take the time to really examine your tongue and you find that it doesn't look good, you're certainly not alone. In many cases, you don't have to be an acupuncturist or Chinese herbalist to recognize that issues are occurring internally. A tongue demonstrating health is pink, flat, and smooth, with a consistent shape and a thin clear coating. But most of us have a tongue that looks very different. Regardless of how dramatic our tongue looks, it's important to know that a well-trained acupuncturist or Chinese herbalist can likely treat the internal causes of your tongue's appearance.

The serious issues with our climate are reflected in the significant imbalances within us physically, mentally, and emotionally. The number of people projected to have cancer and die from the condition is a sobering indication of what's occurring internally. But just as ignoring the data about climate change won't stop the planet from heating, avoiding what's happening to our organs won't promote health.

It takes fortitude to really examine what's happening—both to us as individuals and to the planet. It can be difficult to face the data of climate change: the rates of melting ice sheets and deforestation, the rise of temperatures, and the increase in storms can seem like a continuous stream of bad news. Similarly, when we first start to face what our tongues are telling us about our internal ecology, it can feel overwhelming.

Ultimately, however, it's much harder to attempt to sustain something that's out of balance than it is to maintain something that's moving toward health. While it may feel challenging, especially in the beginning, to pay attention to the state of our organs, there are important opportunities that arise from tongue diagnosis. There is the chance to understand the source of our personal symptoms and to prevent disease before it occurs. And there's the opportunity to realize that what is happening with the climate is indeed happening within us.

Many of us, upon examination, would see that our tongues are red. The amount of heat within us has reached significant and toxic levels, denoted by raised red dots, especially on the backs of our tongues. Our internal heat is a reflection of increasing rates of greenhouse gas emissions and their warming effect on the planet. In particular, it's a mirror of the disproportionate heating and destabilizing effect of the

United States on the planet. As we discussed in chapter 2, the United States has five percent of the world's population but creates about twenty-five percent of global emissions. Just as we are a major cause of a warming planet, many of us have become dramatically overheated as well.

And just as the planet's ability to sequester greenhouse gases and maintain a cool and stable climate has been compromised, many of us have vertical cracks in our tongues indicating our Yin deficiency. As we also discussed in chapter 2, we continue to cut down vast amounts of forests. Permafrost, ice sheets, and glaciers are melting, and enormous amounts of the potent greenhouse gas methane is bubbling from the ocean floor and being released from bogs. All of this indicates a loss of the cooling effects of Yin on a global scale. And on many of our tongues, the crack goes all the way from the back to the front and can be wide and deep, indicating our systemic internal Yin deficiency.

It's also common to see an indentation on the back of the tongue. This personal jing deficiency is a small sign of the larger issue of a culture lacking the wisdom to live sustainably. Many of us have tongues that are deviated and have curled sides, indicating internal wind and Liver Qi stagnation. This excess of Wood is an internal condition that mirrors our country's overemphasis on growth, newness, and conflict. Our internal wind is also mirroring the increasing frequency and force of storms that are the result of our changing climate.

Just as there has been climate change in the world around us, there have been changes to our internal climate. Taking the time to look at our tongues and understand what's happening within us is of vital importance both to our health and the well-being of the planet. One important way to help the planet is to help ourselves. Understanding the balance of our Yin and Yang and the condition of our Five Phases presents the opportunity to recognize that the heat and lack of coolant that climate science is describing globally is happening within us as well. It also allows us to see that our cultural overemphasis on Yang and undervaluing of Yin is affecting us and our organs.

Not only can Chinese medicine help us understand the larger issues of our warming planet, it can also actively promote personal health and address individual symptoms of all kinds. The good news from looking at your tongue is that it gives you an opportunity to understand what is, and is not, promoting your own health. Similarly, the good news of climate change is that it can be our collective wake-up call to examine and address the deeper issues of the crisis. And these two issues—our internal condition and the state of the climate—are merely the same issue occurring on different scales.

CHAPTER 11

The Opportunities of Climate Change

As I sit down to write this final chapter, it's the beginning of September. We've just entered the fifth phase of the year: the harvest season. Harvest is a time of abundance, and here in northwestern Vermont, our local farms are overflowing with food. Corn, squash, tomatoes, melons, kale, eggplants, and zucchini fill the farm stands. Our community-supported agriculture (CSA) share from the small, organic, family-run farm down the road provides us with bags full of these foods each week. Not only are they nourishment for the body—it's also a visual feast. The vibrant hues of yellow, orange, green, purple, and white fill our plates each day.

In our garden, harvest is also at its peak. The tomatoes and greens I wrote about planting over the summer in chapter 8 continue to provide us with an incredible amount of food. Our chard plants are four feet high, each with at least a dozen big, dark-green delicious leaves. Many of our tomato plants don't even seem like plants any more— four feet wide and seven feet tall, having even outgrown their six-foot stakes, they have become bushes, with dozens of small, sweet, orange and red cherry tomatoes hanging from their branches. Some of the tomatoes are so large they completely fill my hand. As I've done for many years, we're growing some of the older, nonhybridized heirloom varieties this year. It's another example of the value of the old and the traditional: the Brandywine and beefsteak tomato plants produce an incredible amount of food. Some of the tomatoes weigh nearly two pounds, and their dense, sweet fruits are one of the great pleasures of gardening in Vermont.

Outside our garden, at the edge of the forest that I look at as I write, the wild plants are telling the same seasonal story. The several varieties of goldenrod that grow at the transition between the trees and the grass are three to four feet tall. As their name indicates, they have large golden-yellow flowers that can be a foot and a half across. I can see whole fields of dense, flowering goldenrods that create acres of bright yellow all around our house. And with all these flowers comes pollen and nectar. The bees and hummingbirds love to visit the goldenrod this time of year, as it's so prolific and one of the few abundant wildflowers in this season.

It's not only all the food that's available and the abundance of flowering wild plants—the sounds in the woods and fields are also telling us it's late summer. The cicadas have been singing nonstop day and night for the past three weeks. They will get quiet if it is too hot and bright during the day, but you can still hear them continuously if you listen closely. When it's cool and dark, especially from sunset to sunrise, the sound of the cicadas is one continuous high-pitched song. At night in our yard, which is surrounded by woods, it's a nearly constant hum that undulates slightly up and down in pitch and volume.

The weather also tells us that it's harvest season. While a few days in the past week have been in the high eighties—hot by Vermont standards—over the last three weeks you can feel that things are mostly starting to cool down. During the day, temperatures are in the high seventies, while at night they're in the high forties to low fifties. With lower temperatures comes increased humidity, and over the last few weeks, outside has been thick with moisture, where you can walk outside and *feel* the air.

These are all signs it's the fifth season—the harvest—which Chinese medicine associates with Earth. This phase is associated with abundance, as is the case in our garden and the farm down the road. The sound of the harvest is singing, as with the continuous chorus of the cicadas. Dampness characterizes the climate of this season, found in the heavy humidity in the air. And the color of the harvest is yellow, which we can see in the abundantly flowering goldenrod.[1]

This is the time of year when we reap what we've sown, at all levels of our lives. Just as the tomato and chard seedlings planted in our gardens have become the fruit and leaves that we're now eating, our beliefs and how we have been living can come to fruition at this time of year.

An important component of Chinese medicine is that this seasonal bounty comes from the descending energy at this time of year. In other words, the decrease in sunlight and temperature allow the harvest to happen. As we talked about in chapter 8, summer is the season of heat, sunshine, fun, and activity—when everything is at its peak. And as we discussed in chapter 6, spring is when things come back to life, and energy expands and rushes upward and outward as we depart the cold and dark of winter.

Without the descent that occurs in the Earth phase, the expanding Qi of spring and summer would never cease. It's the decreasing energy of the harvest that allows the vegetables that are so prolific this time of year to ripen. If we look at the seasons' function from the view of Yin and Yang, the new growth and warmth of spring and summer act as Yang, and the decreased temperature and sunlight of the harvest act as Yin. Just as Yang continues to be vital to our personal and societal health during this era of climate change, the coolness of Yin is also essential to our well-being and the health of our country. The issue, as always, is one of balance.

A Local Tomato or One Shipped across the World

To look more closely at what Chinese medicine's understanding of the Earth phase can tell us about sustainability, let's return to the image of our garden. In addition to the satisfaction of seeing something grow from a small seedling into a huge bush, the tomatoes my wife and I are now eating are both a nutritional feast and a delight to the senses. Each type of tomato has a different size, shape, color, and taste. The cherry, Brandywine, and beefsteak tomatoes all look and taste like tomatoes, of course, but they vary widely in terms of their juiciness, sweetness, and texture. The Brandywine is the rich hue of red wine; the beefsteak is dense and thick, the tomato equivalent of a piece of meat; and the cherry tomatoes are small and sweet.

What allows this richness of color and flavor to build in the tomatoes is the slowing down of the season. If the tomato plants were to continue to grow quickly, as they do in the spring, this ripening would not happen. If they continued to expand, reaching upward toward the hot sun as they do in summer, again, this ripening could not occur. It is the reduced growth and expansion that the cooling temperatures and decreasing sunshine of the Earth phase that allows the tomatoes to become the sweet, succulent delights that they are. And this descent in our own lives is of equal importance.

As we've talked about in earlier chapters, we have a large-scale cultural overemphasis on expansion, and we have structured our economy based on our belief in continuous growth. We're comfortable with waging medical warfare in how we treat conditions like cancer, and we are encouraged to see nature as a place of continuous competition. In addition to this excess of Wood, our country lacks wisdom and is experiencing a significant deficiency of Water. Culturally, this translates to a lack of foresight, exemplified by continuously burning oil even though there are clearly negative consequences. Individually, many of us are also jing-deficient, exemplified by a dip in the back of our tongues.

In both ourselves and our country, too much Wood is also compromising the Metal, which provides lasting inspiration and is our connection to something larger than our individual lives. In an ill-advised effort to keep our economy growing continuously, we're encouraged to value quantity over quality. In other words, to keep the economy constantly growing, we've had to weaken the effects of the Metal—the sense of the sacred—to maintain this unsustainable economic growth.

Given this overstimulated spring energy of Wood, it's not surprising that the way we communicate is also overheated. The quantity of emails, texts, and tweets we send speaks to the condition of our Fire, often creating heat rather than meaningful connections. Knowing that all phases are interconnected, it also comes as no surprise these imbalances in our lives and our culture mean that the Earth phase is affected as well.

If we don't allow our internal temperature to decrease and our internal sunshine to wane, we cannot experience a substantive harvest within our lives. It might seem like the more-is-better, doing-is-better, new-is-better slogans we're encouraged to live by provide us everything we need. But much of it is a hollow shell of real abundance. Having more new things that we don't need and being busy constantly might give a short-lasting sense of having prosperity in our lives. But it is the decline and the slowing down that comes with the harvest that provides us real, long-lasting nourishment.

The difference between an internal harvest that actually nourishes us and one that only feeds us superficially is similar to the difference between a ripe tomato from your garden and one that's shipped thousands of miles to your grocery store. They both look like tomatoes—but a closer look easily uncovers the differences. The tomatoes that are shipped from around the world to your local store are red and round. They're probably smooth on the outside, firm to the touch, and have some moisture in them.

But once you cut it open, the differences between the grocery store variety and the homegrown tomato are clear. While both the local and far-off tomatoes are red, the one that's shipped thousands of miles was likely picked when it was green. Harvesting it before it was ripe might make sense in terms of increasing sales, but the richness of the flavor and the sweetness of the taste are gone. Instead of an inside that is juicy with vibrant shades of red and orange, they're often dried out, with washed-out hues.

This tradeoff—between the juicy, tasty, colorful local tomato and one shipped from a far-off place—is also a good analogy for the trade-off we're encouraged to make with the rest of our lives and the health of the planet. We're led to believe that the personal and societal equivalents of a dry tomato with little flavor and color are all that there is.

We're taught that doing is better than not doing, more is better than less, and new is better than old—all of which speaks to our overemphasis on Yang and devaluing of Yin. We're also told expansion is better than contraction, and that for the economy to be healthy, it needs to grow forever. We're encouraged to believe that medical warfare is the best way to treat serious conditions and that nature is a place of conflict, which speaks to our overemphasis on Wood. This excess of Yang and Wood cannot create health or balance in us, our country, or our culture. Put simply, too much Yang and Wood simply doesn't work—at least not if we're interested in personal health, a satisfying life, and environmental sustainability.

The Good News of Climate Change

As we talked about in chapter 2, there are over twenty-five years of research that clearly indicate that the planet is warming. A vast amount of science from around the world currently continues to show that our planet is rapidly warming and losing its ability to maintain climate stability.

With the stark realities of climate change, there are deep-reaching, potentially life-threatening consequences. When I talk about the ideas and research in this book at conferences and classrooms around the country, one question I'm often asked is how I remain optimistic. Those who ask these questions are likely feeling a range of emotions that come with an honest assessment of the state of our climate: fear, despair, and hopelessness.

While there's nothing necessarily wrong with any of these emotions, it's possible they're coming from our more usual Western, contemporary view. In particular, these feelings can arise from the per-spective that issues of sustainability and unsustainability, and concerns about health and sickness, are in opposition. For example, for many of us, rising global temperatures, increasing greenhouse gas levels, and more severe storms indicate that we are moving away from environ-mental viability toward possible catastrophe.

It is true that more greenhouse gases in the atmosphere mean that the climate will continue to warm. It's also true that massive worldwide deforestation will decrease the amount of greenhouse gases that trees can absorb, again increasing warming. And increased loss of glaciers and ice sheets also means that more sunlight will be absorbed by the soil and water, again contributing to warming. Without diminishing the severity of the condition of the climate, however, it's essential to recognize that this is only a partial view of the climate picture—one that is based on a linear perspective of the world.

The modern Western vantage point tells us that sustainability and unsustainability, and health and sickness, are on different ends of a spectrum. If we get the flu more often or cancer cells continue to grow, this must be telling us that we're moving toward sickness and farther away from health. From this same perspective, the worse the condition of the climate and global warming becomes, the more we're moving away from sustainability and toward unsustainability.

But if we take the same climate data and look at it from an East-ern perspective, a very different picture emerges. If we understand our lives, the climate, and nature from a view of circularity—as with the Yin-Yang and Five Phases circles—we can see that when something reaches its full expression, it eventually transforms and become some-thing else.

For example, when the sunshine and warmth of day has reached its peak, it's at the pinnacle of Yang in the twenty-four-hour cycle. After this peak, it begins to get darker and colder, which is a decrease in Yang and increase in Yin. This increase in cold and dark and decrease in light and warmth eventually becomes night. In other words, when the Yang of day reaches its peak, it's the beginning of the Yin and the eventual transformation into night.

Similarly, from the Five Phases perspective, the peak of each season leads to the next. In spring, when the season has reached its full expression, it then begins the transition into summer. As we talked about in chapter 6, for example, spring is associated with Wood and wind. If we look at the transition between spring and summer, when spring is at its peak, we often have the most severe seasonal winds, which is a harkening to a change into summer. It's the apex of each season that speaks to the coming movement of the next phase of the year.

Looking at our health, the condition of our country, and the state of the climate, the same idea applies: one thing leads to the next. Rather than being contradictory or oppositional, sickness and unsustainability can naturally transform into health and sustainability. Not only is this possible, it's a much easier transition than trying to prevent it from happening.

The current condition of the planet tells us that we're at an extreme point—one marked by a heavy imbalance in our climate. Similarly, our wide-reaching overemphasis on certain parts of our lives also tells us that we are already very far down a particular cultural path. While it's certainly possible to try to avoid addressing the deeper issues of climate change and continue in the direction we're headed, the natural tendency is for things to change.

Our belief in having more, and in newness, expansion, and conflict, have reached such a far extreme that we are now at or very near the peak—and the strong tendency for what happens next is movement into the next phase. This is the natural transition from the Yang of day to the Yin of night, and from spring and Wood into the other seasons and elements.

There are many things that need to be done to heal our warming planet, such as covering our rooftops with solar panels, eating local food, and not buying things we don't need. Our work toward sustainability also includes addressing our often unquestioned beliefs, especially the value of medical warfare, seeing nature as conflict, and believing that the economy can grow forever. And yet, another crucial aspect of treating the sickness of climate change is simply allowing the Qi in our own lives and in our country to follow a more natural, unimpeded trajectory. Just as day becomes night and spring becomes summer, we are in the midst of a natural transition from Yang to Yin. To address climate change, part of what we need to do is not stand in the way—instead, we can allow it to happen.

A major influence that keeps us stuck in our imbalanced lives is the time, money, and Qi we direct toward perpetuating our Wood and Yang excess. From the lens of Chinese medicine, we can see the hundreds of billions of dollars injected into the U.S. economy after the housing collapse of 2008 as a clear example of a systemic attempt to maintain continuous growth.

As noted in chapter 3, the effort spent on advertising is another example of this culture-wide attempt to maintain our excess. The time and money directed toward convincing us that what we have is not enough is one way we are encouraged to cling to our Yang belief in more and newness at the expense of the Yin of satisfaction and contentment.

In our effort to maintain continuous growth, our drink of choice in the United States has become ubiquitous—coffee. We drink so much coffee because it fits so well with this imbalance. It heats us up, amps up our thinking, and overstimulates our emotions and our organs. It's a dietary example of what's happening to us individually and collectively. And the 150 billion cups we drink each year in the United States is a mirror of the dramatic overheating that's occurring globally with climate change.

If we were able to slow down our continuous, frantic efforts to perpetuate the untenable status quo, this would allow the natural tendencies we see in nature to work through us and our institutions.

In addition to the *doing* of Yang, it's vital that we recognize the Yin of *allowing*. Along with all the work that needs to be done to address climate change, an essential remedy for our warming planet is to allow the natural rhythms to continue. The continuous movement of Qi in nature also happens within us and our country. Even though so many of us feel separated from nature, the influences that occur in the world around us fundamentally affect us as well. Just as day naturally becomes night, Yang naturally becomes Yin.

This transition doesn't occur because of our optimism. If you wake up one morning discouraged about the state of the world, it's still going to get dark and cool at night after the warmth and sunshine of day. And whatever season you're in, it will eventually transition into the next season. These changes happen not because of how hopeful we feel, but because there are larger cycles that are constantly in a state of movement. Looking at our culture and our country, the question isn't *if* these transitions will happen, it's *how* and *when*.

Beyond Optimism

Even though the unsustainability of our individual, internal condition and the imbalances of our external, cultural institutions have reached such an advanced state that it's affecting the entire planet, the rhythms of nature continue. The storms now are more severe, sometimes winter doesn't get cold, and sometimes summer can be blazingly hot. Around the planet, there are an increasing number of places that have too much water, resulting in floods, and locales that have too little water, creating droughts. But even with all of these severe, life-threatening results of climate change, the influences of the Five Phases and Yin-Yang are still plain to see—the tendencies in nature that Chinese medicine has been keenly observing and writing about for millennia endure.

An important way to provide context for all of the important work that needs to happen to address climate change is that there is tremendous inertia encouraging us to make the transition. The natural tendency now would be for us, personally and collectively, to move from overemphasizing Yang to valuing Yin. Rather than seeming like an impossible fight against consumerism and its related overemphasis on new and more things, we can take solace in that day eventually becoming night. And rather than railing against an economy based on continuous growth, we can recognize that spring naturally becomes the other seasons.

Historically, Chinese medicine is fond of using water analogies. For example, one common description of meridians, which we talked about in chapter 1 as external pathways connected to internal organs, is that they are rivers of Qi.[2] Thinking about the condition of own lives and the state of our country, the river is clearly flowing in certain directions. Paying attention to the movements of these currents can help us promote lasting personal health and deep-rooted cultural sustainability.

Rather than feeling like we have to fight against it, rest assured that the flow of the river is taking us where we need to go. In the spirit of this water analogy, think about your life as traveling along a river in a canoe. As with any significant change, be it personal or cultural, there can be times when the water gets rough. There will be moments when traveling gets rocky and the canoe feels unstable. There may even be times when the canoe starts to take on water from high waves lapping against the sides or from rain.

Yet even with these difficulties, there can be a real and tangible sense of moving in the right direction. Essential to this feeling of movement is pointing ourselves in the direction of the flow of the river. Maybe the water is rough, maybe some of it is coming into the boat, and maybe the canoe doesn't always feel stable. But it's much easier, less tiring, and more productive to travel the river in the direction of the current, instead of feeling like you have to constantly paddle upstream.

This is not to imply that there isn't work to be done. If you're navigating down a river with its flow, it's still essential that you pay attention to what's ahead of you. You still need to be aware of the rocks, where you are in relation to the riverbanks, and how fast the water is flowing. Even when you're moving downstream with the current, you still need to keep paddling, both to maintain speed and to help control the direction of the boat. And, of course, you need to make sure that the boat is in good shape and doesn't have any holes in it.

Likewise, to balance the climate, there are many things that need to done. Much of it is of urgent importance, and we don't have the luxury of generations or centuries to make the changes—it's a matter of years or a couple of decades. We must address the vital issues related to how we create electricity, source our food, perceive economic health, and approach medicine. But rather than being in opposition to climate health, all of the current data about climate change can actually be leading us back to health.

The transition from the Yang to the Yin is coming. The only real issue is how gracefully we'll make the transition. Not only are these essential changes for the healing of the climate, they're fundamental to our individual health and the well-being of our country and culture.

To turn around a common phrase and add the wisdom of Chinese medicine, the day can be its brightest and hottest before the dark and cool of night.

Notes

1. The Sickness of Climate Change

1. "Governor: Vermont Seeing Worst Flooding in a Century," *Burlington Free Press,* August 29, 2011, www.burlingtonfreepress.com.
2. Paul U. Unschuld, *Medicine in China: A History of Ideas* (Berkeley: University of California Press, 1985), 72.
3. Harriet Beinfield and Efrem Korngold, *Between Heaven and Earth: A Guide to Chinese Medicine* (New York: Ballantine, 1991), 34.
4. Depending on your view and interpretation of Chinese medicine's history, the text dates to the first century BCE, as per Unschuld, *Medicine in China,* 75, or much farther back, to nearly 2800 BCE. The later date for the *Nei Jing* is presented by Jeffrey Yuen and other more traditional Taoist practitioners and teachers of Chinese medicine in unpublished teaching notes. My personal propensity is to accept the older, more traditional understanding of Chinese medicine in general and the historical understanding of the publication of texts in particular. Regardless of the specific date, the text was first published several millennia ago. Even the more recent publication date places the origin of the text over two thousand years. And regardless of the particulars, the *Yellow Emperor's Classic* clearly articulates an understanding that what is happening in nature has an impact on our well-being.
5. Wang Bing, *Yellow Emperor's Canon of Internal Medicine,* trans. Wu Liansheng and Wu Qi (Beijing: China Science and Technology Press, 1997), 18.

6. Zhang Zhong-Jing, *Shang Han Lun: On Cold Damage,* trans. Craig Mitchell, Feng Ye, and Nigel Wiseman (Brookline, MA: Paradigm, 1999).

7. Jian Min Wen and Garry Seifert, trans., *Wen Bing Xue: Warm Disease Theory* (Taos, NM: Paradigm, 2009), 8–9. The *Wen Bing Xue* is considered a modern classic in that it reflects older medical thinking but was published comparatively recently in relation to other seminal works. The School of Heat was more thoroughly developed during the Ming (1368–1644) and Qing (1644–1911) dynasties, although it reflects thinking that clearly connects to the ideas and practices that predate it by many centuries, including the School of Cold.

8. Liansheng and Qi, trans., *Yellow Emperor's Canon,* 28–30.

9. For an introductory discussion of Yin-deficient heat, see Giovanni Maciocia, *The Foundations of Chinese Medicine: A Comprehensive Text for Acupuncturists and Herbalists* (New York: Churchill Livingstone, 1989), 182–83.

10. While the terms *Yin* and *Yang* are not always capitalized, I have chosen to do so to emphasize their fundamental importance in Chinese medical thought. For an in-depth discussion of Yin and Yang in the *Nei Jing,* see Unschuld, *Medicine in China,* 83–89.

11. Lenny Bernstein, R. K. Pachauri, and Andy Reisinger, *Climate Change 2007: Synthesis Report* (Geneva: IPCC, 2007), 30.

12. James Hansen, *Storms of My Grandchildren: The Truth about the Coming Climate Catastrophe and Our Last Chance to Save Humanity* (New York: Bloomsbury, 2009), xv.

2. The Science of Climate Change

1. Lake Champlain Basin Program, "2011 Lake Champlain Flood Update, Draft," May 27, 2011, 7, www.lcbp.org.

2. Fred Pearce, *With Speed and Violence: Why Scientists Fear Tipping Points in Climate Change* (Boston: Beacon Press, 2007), 41–42.

3. Ibid., 3–5.

4. Ian Sample, "The Father of Climate Change," *The Guardian,* June 30, 2005, www.theguardian.com/environment/2005/jun/30/climatechange.climatechange environment2.

5. Bill McKibben, *Eaarth: Making a Life on a Tough New Planet* (New York: Bloomsbury, 2009), xiii.

6. Ibid., 45 and chap. 1.

7. Food and Agriculture Organization of the United Nations, "Global Forest Resource Assessment 2010: Key Findings 2," www.fao.org/forestry/fra/fra2010/en/.

8. Ibid.

9. Ibid., 3.

10. IPCC, "Contribution of Working Group III to the Fourth Assessment Report of the Intergovernmental Panel on Climate Change, 2007," section 1.3.1, note 8, www.ipcc.ch/publications_and_data/ar4/wg3/en/ch1s1-3.html#footnote8.

11. McKibben, *Eaarth*, 43.

12. Jim Robbins, "What's Killing the Great Forests of the American West?," Yale Environment 360, March 15, 2010, http://e360.yale.edu/feature/whats_killing _the_great_forests_of_the_american_west/2252/.

13. Raymond E. Gullison, Peter C. Frumhoff, Josep G. Canadell, Christopher B. Field, Daniel C. Nepstad, Katharine Hayhoe, Roni Avissar, et al., "Tropical Forests and Climate Policy," *Science* 316:5827 (May 18, 2007), 985–86, doi:10.1126 /science.1136163.

14. Marlowe Hood, "Forests Soak Up Third of Fossil Fuel Emissions," *Cosmos,* July 15, 2011, http://fgdfgfd.cosmosmagazine.com/news/forests-soak-third-fossil -fuel-emissions/.

15. Yude Pan, Richard A. Birdsey, Jingyun Fang, Richard Houghton, Pekka E. Kauppi, Werner A. Kurz, Oliver L. Phillips, et al., "A Large and Persistent Carbon Sink in the World's Forests," *Science* 333:6045 (August 19, 2011), 988–93, www.globalcarbonproject.org/global/pdf/pep/Pan.etal.science.Forest_Sink.pdf.

16. Alex Morales, "Oceans Acidifying Fastest in 300 Million Years Due to Emissions," Bloomberg, March 2, 2012, www.bloomberg.com/news/2012-03-01 /oceans-acidifying-fastest-in-300-million-years-due-to-emissions.html; Yale Environment 360, "Ocean Acidifying Faster Than Any Time in 300 Million Years," March 2, 2012, http://e360.yale.edu/digest/ocean_acidifying_faster_than _any_time_in_300_million_years_study_says/3357/. Citing Bärbel Hönisch, Andy Ridgwell, Daniela N. Schmidt, Ellen Thomas, Samantha J. Gibbs, Appy Sluijs, Richard Zeebe, et al., "The Geological Record of Ocean Acidification," *Science* 335:6072 (March 2, 2012), 1058–63, doi:10.1126/science.1208277.

17. Morales, "Oceans Acidifying."

18. Hansen, *Storms,* 80, 82.

19. One example of Chinese medicine's reference to the oceans is in the categorization of the He-Sea acupuncture points. They are located near and along the knee and elbow creases, where the joints flex. As with much of Chinese medicine, there has been and continues to be much discussion about the use and nature of these points. From my clinical experience, the He-Sea points, especially those along the Yin meridians, are very effective at both clearing heat and tonifying Yin, addressing Yin-deficient heat in particular.

20. Summarized by Eliot Barford, "Rising Ocean Acidity Will Exacerbate Global Warming," *Nature,* August 25, 2013, doi:10.1038/nature.2013.13602, citing Katharina D. Six, Silvia Kloster, Tatiana Ilyina, Stephen D. Archer, Kai Zhang, and Ernst Maier-Reimer, "Global Warming Amplified by Reduced Sulphur Fluxes as a Result of Ocean Acidification," *Nature: Climate Change* 3 (2013), 975–78, doi:10.1038/nclimate1981.

21. Pearce, *With Speed and Violence,* 8.

22. Steve Connor, "Exclusive: The Methane Time Bomb," *The Independent,* September 23, 2008, www.independent.co.uk/environment/climate-change/exclusive-the-methane-time-bomb-938932.html.

23. Steve Connor, "Vast Methane 'Plumes' Seen in Arctic Ocean as Sea Ice Retreats," *The Independent,* December 13, 2011, www.independent.co.uk/news/science/vast-methane-plumes-seen-in-arctic-ocean-as-sea-ice-retreats-6276278.html.

24. Steve Connor, "Danger from the Deep: New Climate Threat as Methane Rises from Cracks in Arctic Ice," *The Independent,* April 23, 2012, www.independent.co.uk/news/science/danger-from-the-deep-new-climate-threat-as-methane-rises-from-cracks-in-arctic-ice-7669174.html.

25. Gail Whiteman, Chris Hope, and Peter Wadhams, "Climate Science: Vast Costs of Arctic Change," *Nature* 499 (2013), 401–3, doi:10.1038/499401a.

26. Pearce, *With Speed and Violence,* 78–81.

27. Michael Marshall, "Major Methane Release Is Almost Inevitable," *New Scientist,* February 21, 2013, www.newscientist.com/article/dn23205-major-methane-release-is-almost-inevitable.html#.Uie8frxA975.

28. Edward A. G. Schuur and Benjamin Abbott, "High Risk of Permafrost Thaw," *Nature* 480 (December 2011), 32–33, doi:10.1038/480032a.

29. U.S. Environmental Protection Agency, "Sources of Greenhouse Emissions," www.epa.gov/climatechange/ghgemissions/sources.html.

30. Steve Connor, "Glaciers in Retreat around the World," *The Independent,* December 8, 2011, www.independent.co.uk/environment/climate-change/glaciers-in-retreat-around-the-world-6274223.html.

31. A. Rabatel, B. Francou, A. Soruco, J. Gomez, B. Cáceres, J. L. Ceballos, R. Basantes, M. Vuille, J.-E. Sicart, C. Huggel, et al., "Current State of Glaciers in the Tropical Andes: A Multi-Century Perspective on Glacier Evolution and Climate Change," *The Cryosphere* 7, 81–102, doi:10.5194/tc-7-81-2013.

32. Oliver Duggan, "Report Finds Antarctic Thaw Is Twice as Bad as Anyone Thought," *The Independent,* December 26, 2012, www.independent.co.uk/news/science/report-finds-antarctic-thaw-is-twice-as-bad-as-anyone-thought-8431149.html.

33. Kirk Johnson, "Alaska Looks for Answers in Glacier's Summer Flood Surges," *New York Times*, July 22, 2013, www.nytimes.com/2013/07/23/us/alaska-looks -for-answers-in-glaciers-summer-flood-surges.html?_r=0.

34. U.S. Geological Survey, Northern Rocky Mountain Science Center, "Retreat of Glaciers in Glacier National Park," http://nrmsc.usgs.gov/research/glacier _retreat.htm.

35. Allie Goldstein and Kirsten Howard, "Glacier National Park Prepares for Ice-Free Future," *National Geographic*, September 10, 2013, http://newswatch .nationalgeographic.com/2013/09/10/glacier-national-park-prepares-for-ice -free-future/.

36. Gregory T. Pederson, Lisa J. Graumlich, Daniel B. Fagre, Todd Kipfer, and Clint C. Muhlfeld, "A Century of Climate and Ecosystem Change in Western Montana: What Do Temperature Trends Portend?," *Climate Change* 98:1–2 (January 2010) 135–54, http://link.springer.com/article/10.1007%2Fs10584-009-9642-y.

37. Hansen, *Storms*, 74. As I interpret the last part of Hansen's statement, he is suggesting that the predictions we have from the IPCC, which are forecasting very serious, planet-wide changes, may be significantly understating the magnitude of the effects.

3. The Meaning of Climate Change

1. Chinese medicine talks about the thin and thick fluids in our bodies, the *jin* and *ye*, respectively. For a brief discussion of *jin-ye*, see Maciocia, *Foundations of Chinese Medicine*, 55.

2. The ideas about latency come primarily from my study with Jeffrey Yuen, 88th-generation Taoist priest and internationally recognized teacher and practitioner of Chinese medicine. He does not write extensively, as his tradition, which dates back to the Han dynasty (206 BCE–220 CE) through an unbroken lineage, is primarily an oral one. While we may have a more modern, Western bias toward information that is written down and published, classical Chinese traditions maintain that oral transmission is essential to maintain the contextual nature of the medicine. I began in-depth study with Yuen in 2005 and have numerous references in personal notes from classes on a wide-reaching range of topics about latency in Chinese medical thought and practice.

3. Simon Rogers and Lisa Evans, "World Carbon Dioxide Emissions Data by Country: China Speeds Ahead of the Rest," January 31, 2011, www.guardian.co.uk /news/datablog/2011/jan/31/world-carbon-dioxide-emissions-country -data-co2, citing U.S. Energy Information Administration, "International

Energy Statistics," www.eia.gov/cfapps/ipdbproject/IEDIndex3.cfm?tid=90&pid
=44&aid=8.

4. See Maciocia, *Foundations of Chinese Medicine,* chap. 1, for discussion of Yin
and Yang. My emphasis on how Yin creates Yang is somewhat of an extrapolation beyond what Maciocia discusses, but it does, in my opinion, speak to the
foundational nature of Yin.

5. This is not to imply that I believe that all patients should be treated for the specific
diagnosis of Yin-deficient heat. Rather, I am talking about what I see as a country—the United States—and a culture—of Western, techno-industrialization
—that has been too active for too long to maintain balance.

4. Feeding the Fire of Climate Change

1. Bob Flaws, *The Tao of Healthy Eating: Dietary Wisdom from Chinese Medicine*
(Boulder, CO: Blue Poppy Press, 2001), 54.

2. For more information about dampness, see Beinfield and Korngold, *Between
Heaven and Earth,* 64–66, 144–50; for a discussion of damp heat, pp. 66, 69, 73,
78.

3. For a brief discussion of the term summer heat, see Maciocia, *Foundations of
Chinese Medicine,* 297–98.

4. Flaws, *Tao of Healthy Eating,* 53–55.

5. For some of the many examples of Western medical ideas about coffee, see:
Donald Hensrud, "Is Coffee Good or Bad for Me?," Mayo Clinic, www.mayo
clinic.org/healthy-living/nutrition-and-healthy-eating/expert-answers/coffee
-and-health/faq-20058339; Rob van Dam, "Ask the Expert: Coffee and Health,"
Harvard School of Public Health, www.hsph.harvard.edu/nutritionsource
/coffee/; Neil Osterweil, "Say It's So, Joe: The Potential Health Benefits—and
Drawbacks—of Coffee," Web MD, www.webmd.com/food-recipes/features
/coffee-new-health-food.

6. Dan Bensky and Randall Barolet, *Chinese Herbal Medicine: Formulas and Strate-
gies* (Seattle: Eastland Press, 1990). Rou Gui (Cinnamon Bark) is used in the
Chinese herbal formulas Jin Gui Shen Qi Wan (Kidney Qi Pills from the Golden
Cabinet) and You Gui Wan (Restore the Right) to strengthen the Qi and Yang
of the Kidney. Both formulas continue to be widely used today, to the extent
that they are available in prepared pill form. Although both Kidney Qi Pills and
Restore the Right are considered strengthening and warming—and therefore
Yang—many of the herbs in the formulas are cooling and relaxing—which are
Yin. For example, in the Kidney Qi Pill, Rou Gui and the other herb that warms

the Yang—Fu Zi (Aconite)—together make up only six grams of the eighty-one-gram formula, a total of about 7.5 percent. Restore the Right contains additional herbs that strengthen the Yang, and Yang components make up a higher percentage of the formula, but a major emphasis is also on preserving the Yin. While formulas do exist that strengthen the Yang exclusively without addressing the Yin, they are discussed less and used less often than formulas that do both. See Bensky and Barolet, *Formulas and Strategies,* 275, for information about Kidney Qi Pills from the Golden Cabinet and 278 for Restore the Right.

7. While this image is common in Chinese medicine and Chinese philosophy, it was most clearly presented to me during my initial Chinese herbal medicine training with Thea Elijah during my master's degree training at the Academy for Five Element Acupuncture from 2002 to 2004. I have several references in personal notes about the analogy of the oil lamp and its relevance to the relationship between Yin and Yang.

8. For a discussion of the effects of cold food and iced drinks on digestion, see Flaws, *Tao of Healthy Eating,* chap. 2, 9–24.

9. See Flaws, *Tao of Healthy Eating,* 10–11 for the idea of the Stomach as a cooking pot that needs to maintain a warm temperature to digest effectively.

10. For a discussion of the connection between dairy and dampness, see Flaws, *Tao of Healthy Eating,* 18–20, 30.

11. "Coffee Facts for National Coffee Day," Live Science, September 29, 2011, www.livescience.com/16297-coffee-facts-national-coffee-day-infographic.html. Doing the simple math of 450 million cups of coffee daily for 365 days in a year, that would equal about 165 billion cups, more than the 150 billion mentioned. Although the yearly amount seems low if the daily amount is correct, the 150 billion cups is still a staggering amount of coffee consumed each year.

12. "83% of U.S. Adults Drink Coffee," Food Product Design, April 1, 2013, www.foodproductdesign.com/news/2013/04/83-of-us-adults-drink-coffee.aspx.

13. U.S. Census Bureau, "State & County QuickFacts," http://quickfacts.census.gov/qfd/states/00000.html.

14. For a brief description of the effects of green tea, see Flaws, *Tao of Healthy Eating,* 105. Flaws explains that green tea is cool and a diuretic. In particular, my clinical experience is that its diuretic properties are part of its ability to clear out dampness.

15. For a discussion of Western herbal use of dandelion, see Peter Holmes, *The Energetics of Western Herbs: Treatment Strategies Integrating Western and Oriental Herbal Medicine* (Boulder, CO: Snow Lotus Press, 1998). 677–79. For the

Chinese understanding of dandelion, see Dan Bensky, Andrew Gamble, and Ted J. Kaptchuk, *Chinese Herbal Medicine: Materia Medica* (Seattle: Eastland Press, 1993), 89–90.

16. Bensky et al., *Materia Medica,* 89.

17. Holmes, *Energetics of Western Herbs,* 677–79.

18. For more in-depth descriptions of making decoctions, or strong teas, from plants, see James Green, *The Herbal Medicine-Maker's Handbook: A Home Manual* (Berkeley, CA: Crossing Press, 2000), chap. 9, 111–16. Also see Richo Cech, *Making Plant Medicine* (Williams, OR: Horizon Herbs, 2000), 65–72.

5. The Water Phase

1. Many scholars and practitioners of Chinese medicine believe that "Five Elements" is a mistranslation of the term *wu xing. Wu* can clearly be translated as the number five, and *xing* in my mind does imply more of the idea of *phases* than *elements.* The word *phase* suggests movement and change, while *elements* could suggest something like the periodic table of chemical building blocks of life, such as carbon, hydrogen, oxygen, and so on. This molecular focus is clearly not what Chinese medicine is referring to, which makes the term *Five Phases* a more accurate, though less often used, modern translation. For a more scholarly discussion of the Five Phases, see Unschuld, *Medicine in China,* 52–61. For a discussion of the mistranslation of *wu xing* and a discussion of Five Phases theory, see Ted J. Kaptchuk, *The Web That Has No Weaver: Understanding Chinese Medicine* (New York: McGraw-Hill, 2000), 437, and the remainder of appendix F, respectively.

2. I've chosen to capitalize the phrases Five Elements and Five Phases to emphasize their historical and current importance in Chinese medicine, similar to my capitalization of Yin-Yang.

3. J. R. Worsley's perspective on the Five Elements includes an emphasis on constitutional treatments, where one of the Five Phases is understood to be our source of both strength and weakness. Similar to the constitutional idea in homeopathy, Worsley's interpretation emphasizes that the root cause of both sickness and health is this elemental constitution. See Peter Eckman, *In the Footsteps of the Yellow Emperor: Tracing the History of Traditional Acupuncture* (San Francisco: Long River Press, 2007), for a discussion of Worsley's development of the Five Elements constitutional approach. See also J. R. Worsley, *Traditional Acupuncture: Traditional Diagnosis* (Taos, NM: Redwing, 1990), for his presentation about his ideas on this constitutional Five Elements model.

4. Kaptchuk, *Web That Has No Weaver*, 84.

5. Ibid., 57.

6. For a discussion of jing, see Maciocia, *Foundations of Chinese Medicine*, 38–41.

7. Kaptchuk, *Web That Has No Weaver*, 55–56. I am focusing on the concept of prenatal essence in this discussion, and not including the idea of postnatal essence that can be created from the food we eat and the air we breathe. The idea of the prenatal jing we get from both parents and share with our children seems most relevant to the discussion of the underlying causes and consequences of climate change.

8. See ibid., 55–56, 83–87 for additional discussion.

9. Personal notes from master's degree acupuncture training with Thea Elijah, 2002–2004.

10. See Richard Heinberg, *The Party's Over: Oil, War and the Fate of Industrial Societies,* 2nd ed. (Philadelphia: New Society, 2005), 57–117, for a discussion of the history and current use of oil.

11. For a more in-depth discussion of peak oil, see: Heinberg, *Party's Over;* Kenneth S. Deffeyes, *Beyond Oil: The View from Hubbert's Peak* (New York: Hill and Wang, 2006); and James Howard Kunstler, *The Long Emergency: Surviving the End of Oil, Climate Change, and Other Converging Catastrophes of the Twenty-First Century* (New York: Grove Press, 2006).

12. For a discussion of tar sands oil in Canada, see the Samuel Avery, *The Pipeline and the Paradigm: Keystone XL, Tar Sands, and the Battle to Defuse the Carbon Bomb* (Washington, DC: Ruka Press, 2013), 1–13. Also see Andrew Nikiforuk, *Tar Sands: Dirty Oil and the Future of a Continent* (Vancouver: Greystone Books, 2010), chap. 2, 11–17.

13. McKibben, *Eaarth,* 28–30.

14. My perspective of jing as including a generational understanding of environmental sustainability is an expansion of what is presented in current Chinese medicine by Maciocia in *Foundations of Chinese Medicine* and Beinfield and Korngold in *Between Heaven and Earth,* for example. I do believe that it is a reasonable extrapolation of the ideas of our individual jing to a global scale.

15. For information about these jing tonics, see Bensky et al., *Materia Medica.*

16. Chinese herbs like Tu Su Zi (*Cuscuta* Seed), Shu Di Huang (Prepared *Rehmannia* Root), Gou Qi Zi (*Lycium* Berries), and He Shou Wu (*Polygonum* Root) have been recognized for generations to tonify jing. The formula Qi Bao Mai Ren Dan, which has the poetic translation Seven-Treasure Special Pill for Beautiful Whiskers, tonifies jing as well. See Bensky et al., *Formulas and Strategies,* 273–74.

6. The Consequences of Continuous Growth

1. For a discussion of the Sheng cycle, see Beinfield and Korngold, *Between Heaven and Earth,* 95–99.

2. Liansheng and Qi, trans., *Yellow Emperor's Canon,* 64–68, and ibid., 437–40.

3. While there is a place for newness even in our era of yang and Wood excess, climate change makes it clear that we clearly lack the wisdom to create new life forms. As with our warming planet, bioengineering and genetic modification are experiments in which we might not know the exact outcome but can certainly anticipate the general results. Like climate change, changing life at such a basic genetic level without first attempting to understand the long-term consequences to the planet or our lives is another indication of a lack of cultural wisdom.

4. For more about the idea of wind in Chinese medicine, see Kaptchuk, *Web That Has No Weaver,* 3, 150–52, and Beinfield and Korngold, *Between Heaven and Earth,* 64, 66, 70, 167, 169.

5. For a nice synopsis of the differences in Eastern and Western medical perspectives, see Beinfield and Korngold, *Between Heaven and Earth,* chaps. 2–3, 17–45.

6. Unschuld, *Medicine in China,* 31.

7. For the reference to "all against all," see ibid., 37. For a discussion of the simultaneous development of a cultural, military, and medical view during the Warring States era of continuous conflict, see ibid., 29–46.

8. Looking at the assumptions of modern Western biology from an older Eastern perspective does not in any way invalidate or diminish the importance of climate research. Understanding what is happening biologically is of vital importance to addressing both the more superficial and deeper issues of our warming planet. I see the emphasis on competition in nature as one interpretation among many.

7. Quality Controls Quantity

1. For more in-depth discussion of K'o cycle, see Beinfield and Korngold, *Between Heaven and Earth,* 95–103, and Maciocia, *Foundations of Chinese Medicine,* 19, 21–23.

2. Yin and Yang are not often used to describe the K'o and Sheng cycle, including the contemporary J. R. Worsley inspired interpretations of the Five Elements. I do think it an accurate blending of the ideas of Yin-Yang and the Five Phases to say that the Sheng cycle is Yang and the K'o cycle Yin. For a brief discussion of a

historical perspective on the connection between the Five Phases and Yin-Yang, see Kaptchuk, *Web That Has No Weaver*, 440–41.

3. For more discussion of the Metal, see Beinfield and Korngold, *Between Heaven and Earth*, 204–18.

4. For a brief discussion of Heavenly Qi, see Maciocia, *Foundations of Chinese Medicine*, 83. I have notes from classes with Jeffrey Yuen on several Chinese medicine topics about the function of the Lung, including its role of taking in Ancestral Qi with inhalation.

5. For more information about the Large Intestine, see Kaptchuk, *Web That Has No Weaver*, 96, and Maciocia, *Foundations of Chinese Medicine*, 115.

6. For more information about the functions of the inhalation and exhalation of the Lung, see Kaptchuk, *Web That Has No Weaver*, 90–93, and Maciocia, *Foundations of Chinese Medicine*, 83–84.

7. From notes from master's degree acupuncture training with Thea Elijah, 2002–2004.

8. From many years of study of Tai Chi Ch'uan, I've heard Wolfe Lowenthal use the "pulling silk" metaphor many times. My understanding is that he is sharing what he learned from his teacher Cheng Man Ch'ing, creator of the Yang-style short form.

9. For a discussion of the relationship between Metal and Wood and Water, see, Beinfield and Korngold, *Between Heaven and Earth*, 208–11; Kaptchuk, *Web That Has No Weaver*, 444–46; and Maciocia, *Foundations of Chinese Medicine*, 19–20.

10. Jeffrey Yuen teaches about the five branches of Chinese medicine—acupuncture, herbal medicine, nutrition, massage, and internal practices like Tai Chi and Qi Gong. From the historical development of these branches, each was understood to be a complete medical system unto itself, capable of treating all diseases and promoting individual health. There is a contemporary school of Chinese medicine in California called Five Branches University that incorporates these different aspects of Chinese medicine in the curriculum. A major part of Chinese medical diagnosis and treatment can be seen as balancing the influences of nature within us. The importance of the effects of the Five Phases of the seasons and the weather (hot, cold, damp, dry, etc.) speak to its nature-based perspective. One prominent lineage of Chinese medicine that emphasized the importance of interacting and living in balance with nature was the Naturalist School. While I read the totality of the *Nei Jing* as a medical and philosophical discussion of a nature-based perspective of personal and societal

health, for a very brief discussion of this Naturalist School, see Maoshing Ni, trans., *The Yellow Emperor's Classic of Medicine: A New Translation of the Neijing Suwen with Commentary* (Boston: Shambhala, 1995), 4. Also see chaps. 2, 9, 19, 22, 42, 43, and 66 of this translation for discussion about the effects of the seasons and the weather on our health and specific treatments to address these conditions.

8. The Dissatisfaction of Too Much Fire

1. There has been a long historical discussion about whether the Pericardium is its own separate organ and whether the Triple Heater is an organ or a function. I'm presenting the perspective that they are in fact separate organs with separate functions. For a brief discussion of the debate about the nature of the Pericardium and Triple Heater, see Maciocia, *Foundations of Chinese Medicine,* chap. 11, 103–4, 117–20.

2. For additional information about the Five Phases correspondences, see Liansheng and Qi, trans., *Yellow Emperor's Canon,* 64–68, and Beinfield and Korngold, *Between Heaven and Earth,* 176–79. For a discussion of the functions of the four Fire organs, see Maciocia, *Foundations of Chinese Medicine,* 71–76 for the Heart, pp. 103–4 for the Pericardium, p. 114 for the Small Intestine, and pp. 117–20 for the Small Intestine.

3. This discussion about the different scales of relationships associated with the Pericardium and Triple Heater is not written about in this way in any of the Chinese medicine texts cited. It is my extrapolation on the Yin and Yang aspects of the Fire as embodied by the Pericardium and Triple Heater, respectively. For example, as a Yin organ of the Fire, the Pericardium is about our relationships to smaller groups, while the Triple Heater is the Yang organ and is about our interaction with larger groups and the world.

4. For a brief discussion of Heart-Kidney communication, see Maciocia, *Foundations of Chinese Medicine,* 23–24. Also see Lonny S. Jarrett, *The Clinical Practice of Chinese Medicine* (Stockbridge, MA: Spirit Path Press, 2003), 5–10, as well as pp. 10–13 for commentary on the connection between the Heart-Kidney axis and cultural changes.

9. Cancer and Climate Change

1. American Cancer Society, "Cancer Facts and Figures: 2014," www.cancer.org /acs/groups/content/@research/documents/webcontent/acspc-042151.pdf;

American Cancer Society, "Lifetime Probability of Developing or Dying From Cancer," www.cancer.org/cancer/cancerbasics/lifetime-probability-of -developing-or-dying-from-cancer.

2. American Cancer Society, "Cancer Basics," www.cancer.org/cancer/cancerbasics /what-is-cancer.

3. American Cancer Society, "Cancer Facts and Figures," 14.

4. American Cancer Society, "Lifetime Probability."

5. American Cancer Society, "What Causes Cancer," www.cancer.org/cancer/cancer causes/index.

6. See Flaws, *Tao of Healthy Eating*, 71, for the clearest presentation of the temperature and nature of alcohol.

7. Maciocia attributes this description of tobacco to Qu Ci Shan. See Giovanni Maciocia, "The View of Tobacco in Chinese Medicine," blog post, February 3, 2009, http://maciociaonline.blogspot.com/2009/02/view-of-tobacco-in-chinese -medicine.html.

8. Different cultural perspectives have different views on the use of tobacco. Many Indigenous people here in the United States and worldwide consider it a sacred plant and continue to use it ceremonially. In no way am I criticizing this, and I strongly believe that the old and traditional understandings of Indigenous people are essential to our personal health and cultural sustainability. My intention with the discussion of a Chinese medicine understanding of the nature of tobacco is to present its hot, stimulating, potentially toxic nature in the context of our overstimulated lives and overheating planet. I have no doubt that within the context of Indigenous cultural and spiritual beliefs, the use of tobacco can be important tools for fostering health and well-being.

9. As with the discussion of coffee, I hope the brief presentation about ideas in Chinese medicine is not read as a condemnation of people who smoke or use tobacco. It's not about morality or shame; it's about temperature. We live in a hot, overstimulated, jing- and wisdom-deficient culture. So many of us are drawn to the stimulation of hot substances like coffee and tobacco because it resonates with how we're encouraged to live and what we're encouraged to believe. The Yang "up" we get from both substances mirrors the Yang quest for continuous economic growth, newness, and doing. As with coffee, if you'd like help to stop smoking, a well-trained Chinese medicine practitioner should be able to help. In my clinical experience, treating many people who have stopped smoking, Chinese medicine can significantly reduce cravings and help clear out the heat and moisten the dryness that has very likely been created. If needed, it can also tonify the jing.

10. See Flaws, *Tao of Healthy Eating*, 71–109, for a list of foods and their temperatures, including that of the meats discussed.

11. For a discussion of the use of the Eight Extraordinary Meridians to treat RNA, DNA, and cancer, see Ann Cecil-Sterman, *Advanced Acupuncture: A Clinical Manual* (New York: Classical Wellness Press, 2012), 220–21. For the specific uses of the Eight Extras, see 218–336.

12. For a discussion of the nature of Centipede and Scorpion and current and historical uses of both substances to strongly treat wind, heat, and toxicity, see Bensky et al., *Materia Medica*, 427–28.

13. Reference to the first written source of the text comes in the publisher's forward, written by Bob Flaws, in Yang Shou-zhong, trans., *The Divine Farmer's Materia Medica: A Translation of the Shen Nong Ben Cao Jing* (Boulder, CO: Blue Poppy Press, 1998), iii. As mentioned earlier in the endnote discussion about the date of the *Yellow Emperor's Classic,* the date for the *Divine Farmer's Materia Medica* is also open to debate. As noted by Flaws, the first reference dates to the Qin dynasty (221–216 BCE).

14. See Bensky et al., *Materia Medica*, 323–24, for discussion of the effects and nature of Gan Cao (Licorice).

15. See ibid., 68–70, for discussion of the effects and nature of Sheng Di Huang (Raw *Rehmannia*) and Xuan Shen *(Scrophularia).*

16. The need to balance the invigorating effects of Yang with the cooling of Yin is one of the many benefits of internal practices like Tai Chi and Qi Gong. As forms of exercise, they are slow-moving and contemplative, indicating their internal, Yin focus. For most people most of the time, they create little if any sweating. And when sweating does occur, as can happen when practicing in a hot environment, it's certainly not the intended purpose of the practice. Rather than dispersing Qi and sweating out our Yin, as can happen with more Yang, externally focused exercise, Tai Chi and Qi Gong can strengthen our energy and increase our fluids.

17. National Cancer Institute at the National Institutes of Health, "Cancer Costs Projected to Reach at Least $158 Billion In 2020," January 1, 2011, www.cancer.gov/newscenter/newsfromnci/2011/CostCancer2020, citing Angela B. Mariotto, K. Robin Yabroff, Yongwu Shao, Eric J. Feuer, and Martin L. Brown, "Projections of the Cost of Cancer Care in the United States: 2010–2020," *Journal of the National Cancer Institute* 103:2 (January 19, 2011), 117–28, doi:10.1093/jnci/djq495.

18. American Cancer Society, "Treatments Linked to the Development of Second Cancers," January 30, 2012, www.cancer.org/cancer/cancercauses/othercarcinogens

/medicaltreatments/secondcancerscausedbycancertreatment/second-cancers
-caused-by-cancer-treatment-treatments-linked-to-second-cancers.

19. Ibid.

20. As with other patient case studies, his name has been change to protect confi-
dentiality. For the sake of privacy, the specific numbers of his blood work have
also been changed. However, the general relationship between the changes in
his blood work, as well as the specific percentage of change in his white blood
cell count, accurately reflect the differences recorded by Western lab tests.

21. Medline Plus, "Chronic Lymphocytic Leukemia (CLL)," March 23, 2014,
www.nlm.nih.gov/medlineplus/ency/article/000532.htm.

22. Ben being told to wait until his white blood cells had doubled, justifying
chemotherapy, is an individual example of the more systemic issue in Western
medicine of waiting until the disease has appeared before it's treated.

23. The *Wen Bing Xue* is discussed in chap. 1. See Wen and Seifert, *Wen Bing Xue*,
36–38. Also see Maciocia, *Foundations of Chinese Medicine*, 194, for a very brief
discussion.

24. The formula was a modification of Xi Jiao Di Huang Tang (Rhinoceros Horn)
and *Rehmannia* Decoction. In place of the horn from the endangered rhinoceros, Xuan Shen *(Scrophularia)* makes an excellent substitute. Along with the
additions of Quan Xie (Scorpion) and Wu Gong (Centipede), Da Huang (Rhu-
barb Root) was used to clear out blood heat and toxins through the bowels.

25. Tait D. Shanafelt, Heidi Gunderson, and Timothy G. Call, "Commentary:
Chronic Lymphocytic Leukemia—The Price of Progress," *The Oncologist
Express*, May 12, 2010, doi:10.1634/theoncologist.2010-0090.

10. Internal Climate Change

1. Barbara Kirschbaum, *Atlas of Chinese Tongue Diagnosis* (Seattle: Eastland Press,
2000). See Dominique Hertzer's Foreword, xii–xiv, for specific references, and
the entire forward for a brief discussion of the history of tongue diagnosis.

2. The Eastern and Western perspectives on the importance of the brain speak
to the different cultural views that have shaped the medical traditions. With
Chinese medicine the brain is described as a curious organ that is accessed directly through treating the Gallbladder. For a brief discussion about the unusual
qualities of the Gallbladder, see Maciocia, *Foundations of Chinese Medicine*,
115–16. However, Maciocia does not use the word *curious* to describe the brain.
My categorization of the brain as a curious organ comes from my studies with
Jeffery Yuen. The Wai Ke tradition in Chinese medicine specializes in treating

neurological conditions and associates their root causes primarily to an excess of internal wind. As discussed earlier, wind is associated with the Wood phase as well as the Gallbladder and Liver. Interestingly, the Wai Ke tradition also understands all dermatological conditions as coming primarily from wind. As a result, practitioners working within this perspective specialize in treating both neurological and dermatological conditions. At our clinic, we regularly utilize the Wai Ke diagnostic and treatment principles to address neurological and skin conditions, and we have consistently seen excellent results. In addition to the case studies from chap. 6, on Wood, about grand mal seizures and migraines, we've seen dramatic improvements and complete resolution with epileptic seizures, childhood seizures, and febrile seizures. We've also seen excellent results addressing the causes of psoriasis, eczema, acne, and a very wide range of other skin conditions, including skin cancers. My understanding and clinical application of the Wai Ke lineage also comes from my training with Jeffery Yuen. There are no written references to it that I'm aware of that have been translated into English.

3. Most of the discussion in this chapter corresponds to what is currently written about tongue diagnosis, including in-depth clinical texts like Kirschbaum, *Chinese Tongue Diagnosis,* as well as more introductory texts like Beinfield and Korngold, *Between Heaven and Earth.* Kirschbaum presents a very concise visual differentiation of some of the major historical differences in the organs associated with parts of the tongue (p. 2). The view I'm presenting this chapter is very similar to fig. 5, p. 2, with the difference being that I don't associate the condition of the two Intestines with different sides of the back of the tongue.

4. The differentiation between vertical and horizontal cracks comes from my study with Jeffrey Yuen. For both Kirschbaum, *Chinese Tongue Diagnosis,* and Beinfield and Korngold, in *Between Heaven and Earth,* cracks in the tongue are associated with dryness and Yin deficiency.

5. A puffy tongue can also indicate dampness, where too much fluid is accumulating. From my clinical experience, one difference between the puffiness of Qi deficiency and dampness is that with a lack of energy, the enlarged size of the tongue can appear as spread out rather than concentrated in a more clearly defined area. As part of this difference, when the top of the tongue associated with a particular organ is puffy, I usually associate this with dampness rather than Qi deficiency.

6. For a condensed discussion of the function of the Spleen and Stomach, see Maciocia, *Foundations of Chinese Medicine,* 89–92. Also see pp. 111–13, citing the

classical Five Phases text, the *Nei Jing*. To read a primary source presentation of the function of both organs, see Ni, *Yellow Emperor's Classic:* for a discussion of the Spleen, see pp. 21, 42, 289, and 314. For a discussion of the Stomach, see pp. 40–41, 46–47, 89, and 314. See pp. 116–17 for a discussion of the interrelation of these organs.

7. See Maciocia, *Foundations of Chinese Medicine,* 78–79, for a brief discussion of the role of the Liver in ensuring the smooth flow of Qi.

8. See chap. 10, note 2, above for more information about the Wai Ke tradition of Chinese medicine and its focus on treating internal wind and neurological conditions of all kinds.

9. Kidney Qi and Yang deficiency are also associated with symptoms of the lower back, legs, knees, feet, and ankles. In my clinical experience, osteoarthritis is often associated with Kidney Yin-deficient heat as well as deficient jing. Because it is a basis for the Qi, Yin, and Yang of the Kidney, when there's jing deficiency, all of these are deficient as well. See Maciocia, *Foundations of Chinese Medicine,* 252–57, for a brief discussing of the differences in deficient Qi, Yin, and Yang of the Kidney as well as jing deficiency.

11. The Opportunities of Climate Change

1. See Maciocia, *Foundations of Chinese Medicine,* 21, for a brief presentation about the association of the Earth phase. For a more in-depth discussion of Earth, see Beinfield and Korngold, *Between Heaven and Earth,* 190–203.

2. See Beinfield and Korngold, *Between Heaven and Earth,* 236–37, for a brief discussion.

Bibliography

American Cancer Society. "Cancer Basics." www.cancer.org/cancer/cancerbasics
/what-is-cancer.
———. "Cancer Facts and Figures: 2014." www.cancer.org/acs/groups/content
/@research/documents/webcontent/acspc-042151.pdf.
———. "Lifetime Probability of Developing or Dying From Cancer." www.cancer
.org/cancer/cancerbasics/lifetime-probability-of-developing-or-dying-from
-cancer.
———. "Treatments Linked to the Development of Second Cancers." January 30,
2012. www.cancer.org/cancer/cancercauses/othercarcinogens/medicaltreatments
/secondcancerscausedbycancertreatment/second-cancers-caused-by-cancer
-treatment-treatments-linked-to second-cancers.
———. "What Causes Cancer." www.cancer.org/cancer/cancercauses/index.
Avery, Samuel. *The Pipeline and the Paradigm: Keystone XL, Tar Sands, and the Battle
to Defuse the Carbon Bomb.* Washington, DC: Ruka Press, 2013.
Barford, Eliot. "Rising Ocean Acidity Will Exacerbate Global Warming." *Nature.*
August 25, 2013. doi:10.1038/nature.2013.13602.
Beinfield, Harriet, and Efrem Korngold. *Between Heaven and Earth: A Guide to
Chinese Medicine.* New York: Ballantine, 1991.
Bensky, Dan, and Randall Barolet. *Chinese Herbal Medicine: Formulas and Strategies.*
Seattle: Eastland Press, 1990.
Bensky, Dan, Andrew Gamble, and Ted J. Kaptchuk. *Chinese Herbal Medicine:
Materia Medica.* Seattle: Eastland Press, 1993.
Bernstein, Lenny, R. K. Pachauri, and Andy Reisinger. *Climate Change 2007: Synthesis
Report.* Geneva: IPCC, 2007.

Bing, Wang. *Yellow Emperor's Canon of Internal Medicine.* Trans. Wu Liansheng and Wu Qi. Beijing: China Science and Technology Press, 1997.

Burlington Free Press. "Governor: Vermont Seeing Worst Flooding in a Century." August 29, 2011. www.BurlingtonFreePress.com.

Cech, Richo. *Making Plant Medicine.* Williams, OR: Horizon Herbs, 2000.

Cecil-Sterman, Ann. *Advanced Acupuncture: A Clinical Manual.* New York: Classical Wellness Press, 2012.

Connor, Steve. "Danger from the Deep: New Climate Threat as Methane Rises from Cracks in Arctic Ice." *The Independent.* April 23, 2012. www.independent.co.uk /news/science/danger-from-the-deep-new-climate-threat-as-methane-rises -from-cracks-in-arctic-ice-7669174.html.

———. "Exclusive: The Methane Time Bomb." *The Independent.* September 23, 2008. www.independent.co.uk/environment/climate-change/exclusive-the -methane-time-bomb-938932.html.

———. "Glaciers in Retreat around the World." *The Independent.* December 8, 2011. www.independent.co.uk/environment/climate-change/glaciers-in-retreat -around-the-world-6274223.html.

———. "Vast Methane 'Plumes' Seen in Arctic Ocean as Sea Ice Retreats." *The Independent.* December 13, 2011. www.independent.co.uk/news/science /vast-methane-plumes-seen-in-arctic-ocean-as-sea-ice-retreats-6276278.html.

Deffeyes, Kenneth S. *Beyond Oil: The View from Hubbert's Peak.* New York: Hill and Wang, 2006.

Duggan, Oliver. "Report Finds Antarctic Thaw Is Twice as Bad as Anyone Thought." *The Independent.* December 26, 2012. www.independent.co.uk/news/science /report-finds-antarctic-thaw-is-twice-as-bad-as-anyone-thought-8431149.html.

Eckman, Peter. *In the Footsteps of the Yellow Emperor: Tracing the History of Traditional Acupuncture.* San Francisco: Long River Press, 2007.

Flaws, Bob. *The Tao of Healthy Eating: Dietary Wisdom from Chinese Medicine.* Boulder, CO: Blue Poppy Press, 2001.

Food and Agriculture Organization of the United Nations. "Global Forest Resource Assessment 2010: Key Findings 2." www.fao.org/forestry/fra/fra2010/en/.

Food Product Design. "83% of U.S. Adults Drink Coffee." April 1, 2013. www.food productdesign.com/news/2013/04/83-of-us-adults-drink-coffee.aspx.

Goldstein, Allie, and Kirsten Howard. "Glacier National Park Prepares for Ice-Free Future." *National Geographic.* September 10, 2013. http://newswatch.national geographic.com/2013/09/10/glacier-national-park-prepares-for-ice-free-future/.

Green, James. *The Herbal Medicine-Maker's Handbook: A Home Manual.* Berkeley, CA: Crossing Press, 2000.

Gullison, Raymond E., Peter C. Frumhoff, Josep G. Canadell, Christopher B. Field, Daniel C. Nepstad, Katharine Hayhoe, Roni Avissar, Lisa M. Curran, Pierre

Friedlingstein, Chris D. Jones, Carlos Nobre, et al. "Tropical Forests and Climate Policy." *Science* 316:5827 (May 18, 2007), 985–86. doi:10.1126/science.1136163.

Hansen, James. *Storms of My Grandchildren: The Truth about the Coming Climate Catastrophe and Our Last Chance to Save Humanity.* New York: Bloomsbury, 2009.

Heinberg, Richard. *The Party's Over: Oil, War and the Fate of Industrial Societies.* 2nd ed. Philadelphia: New Society, 2005.

Hensrud, Donald. "Is Coffee Good or Bad for Me?" Mayo Clinic. www.mayoclinic. org/healthy-living/nutrition-and-healthy-eating/expert-answers/coffee-and -health/faq-20058339.

Holmes, Peter. *The Energetics of Western Herbs: Treatment Strategies Integrating Western and Oriental Herbal Medicine.* Boulder, CO: Snow Lotus Press, 1998.

Hönisch, Bärbel, Andy Ridgwell, Daniela N. Schmidt, Ellen Thomas, Samantha J. Gibbs, Appy Sluijs, Richard Zeebe, Lee Kump, Rowan C. Martindale, Sarah E. Greene, et al. "The Geological Record of Ocean Acidification." *Science* 335:6072 (March 2, 2012), 1058–63. doi:10.1126/science.1208277.

Hood, Marlowe. "Forests Soak Up Third of Fossil Fuel Emissions." *Cosmos.* July 15, 2011. http://fgdfgfd.cosmosmagazine.com/news/forests-soak-third-fossil-fuel -emissions/.

Intergovernmental Panel on Climate Change. "Contribution of Working Group III to the Fourth Assessment Report of the Intergovernmental Panel on Climate Change, 2007." Section 1.3.1, note 8. www.ipcc.ch/publications_and_data/ar4 /wg3/en/ch1s1-3.html#footnote8.

Jarrett, Lonny S. *The Clinical Practice of Chinese Medicine.* Stockbridge, MA: Spirit Path Press, 2003.

Johnson, Kirk. "Alaska Looks for Answers in Glacier's Summer Flood Surges." *New York Times.* July 22, 2013. www.nytimes.com/2013/07/23/us/alaska-looks-for -answers-in-glaciers-summer flood-surges.html?_r=0.

Kaptchuk, Ted J. *The Web That Has No Weaver: Understanding Chinese Medicine.* New York: McGraw-Hill, 2000.

Kirschbaum, Barbara. *Atlas of Chinese Tongue Diagnosis.* Seattle: Eastland Press, 2000.

Kunstler, James Howard. *The Long Emergency: Surviving the End of Oil, Climate Change, and Other Converging Catastrophes of the Twenty-First Century.* New York: Grove Press, 2006.

Lake Champlain Basin Program. "2011 Lake Champlain Flood Update, Draft." May 27, 2011. www.lcbp.org.

Live Science. "Coffee Facts for National Coffee Day." September 29, 2011. www.livescience.com/16297-coffee-facts-national-coffee-day-infographic.html.

Maciocia, Giovanni. *The Foundations of Chinese Medicine: A Comprehensive Text for Acupuncturists and Herbalists.* New York: Churchill Livingstone, 1989.

———. "The View of Tobacco in Chinese Medicine." Blog post. February 3, 2009. http://maciociaonline.blogspot.com/2009/02/view-of-tobacco-in-chinese -medicine.html.

Mariotto, Angela B., K. Robin Yabroff, Yongwu Shao, Eric J. Feuer, and Martin L. Brown, "Projections of the Cost of Cancer Care in the United States: 2010–2020." *Journal of the National Cancer Institute* 103:2 (January 19, 2011), 117–28. doi:10.1093/jnci/djq495.

Marshall, Michael. "Major Methane Release Is Almost Inevitable." *New Scientist.* February 21, 2013. www.newscientist.com/article/dn23205-major-methane -release-is-almost-inevitable.html#.Uie8frxA975.

McKibben, Bill. *Eaarth: Making a Life on a Tough New Planet.* New York: Bloomsbury, 2009.

Medline Plus. "Chronic Lymphocytic Leukemia (CLL)." March 23, 2014. www.nlm .nih.gov/medlineplus/ency/article/000532.htm.

Morales, Alex. "Oceans Acidifying Fastest in 300 Million Years Due to Emissions." Bloomberg. March 2, 2012. www.bloomberg.com/news/2012-03-01/oceans -acidifying-fastest-in-300-million-years-due-to-emissions.html.

National Cancer Institute at the National Institutes of Health. "Cancer Costs Projected to Reach at Least $158 Billion In 2020." January 1, 2011. www.cancer.gov /newscenter/newsfromnci/2011/CostCancer2020.

Ni, Maoshing, trans. *The Yellow Emperor's Classic of Medicine: A New Translation of the Neijing Suwen with Commentary.* Boston: Shambhala, 1995.

Nikiforuk, Andrew. *Tar Sands: Dirty Oil and the Future of a Continent.* Vancouver: Greystone Books, 2010.

Osterweil, Neil. "Say It's So, Joe: The Potential Health Benefits—and Drawbacks—of Coffee." Web MD. www.webmd.com/food-recipes/features/coffee-new-health -food.

Pan, Yude, Richard A. Birdsey, Jingyun Fang, Richard Houghton, Pekka E. Kauppi, Werner A. Kurz, Oliver L. Phillips, Anatoly Shvidenko, Simon L. Lewis, Josep G. Canadell, et al. "A Large and Persistent Carbon Sink in the World's Forests." *Science* 333:6045 (August 19, 2011), 988–93. www.globalcarbonproject.org/global /pdf/pep/Pan.etal.science.Forest_Sink.pdf.

Pearce, Fred. *With Speed and Violence: Why Scientists Fear Tipping Points in Climate Change.* Boston: Beacon Press, 2007.

Pederson, Gregory T., Lisa J. Graumlich, Daniel B. Fagre, Todd Kipfer, and Clint C. Muhlfeld. "A Century of Climate and Ecosystem Change in Western Montana: What Do Temperature Trends Portend?" *Climate Change* 98:1–2 (January 2010) 135–54. http://link.springer.com/article/10.1007%2Fs10584-009-9642-y.

Rabatel, Antoine, B. Francou, A. Soruco, J. Gomez, B. Cáceres, J. L. Ceballos, R. Basantes, M. Vuille, J.-E. Sicart, C. Huggel, M. Scheel, Y. Lejeune, Y. Arnaud, et al. "Current State of Glaciers in the Tropical Andes: A Multi-Century Perspective on Glacier Evolution and Climate Change." *The Cryosphere* 7, 81–102. doi:10.5194/tc-7-81-2013.

Robbins, Jim. "What's Killing the Great Forests of the American West?" Yale Environment 360. March 15, 2010. http://e360.yale.edu/feature/whats_killing_the_great_forests_of_the_american_west/2252.

Rogers, Simon, and Lisa Evans. "World Carbon Dioxide Emissions Data by Country: China Speeds Ahead of the Rest." January 31, 2011. www.guardian.co.uk/news/datablog/2011/jan/31/world-carbon-dioxide-emissions-country-data-co2.

Sample, Ian. "The Father of Climate Change." *The Guardian.* June 30, 2005. www.theguardian.com/environment/2005/jun/30/climatechange.climate changeenvironment2.

Schuur, Edward A. G., and Benjamin Abbott. "High Risk of Permafrost Thaw." *Nature* 480 (December 2011), 32–33. doi:10.1038/480032a.

Shanafelt, Tait D., Heidi Gunderson, and Timothy G. Call. "Commentary: Chronic Lymphocytic Leukemia—The Price of Progress." *The Oncologist Express.* May 12, 2010. doi:10.1634/thconcologist.2010-0090.

Shou-zhong, Yang, trans. *The Divine Farmer's Materia Medica: A Translation of the Shen Nong Ben Cao Jing.* Boulder, CO: Blue Poppy Press, 1998.

Six, Katharina D., Silvia Kloster, Tatiana Ilyina, Stephen D. Archer, Kai Zhang, and Ernst Maier-Reimer. "Global Warming Amplified by Reduced Sulphur Fluxes as a Result of Ocean Acidification." *Nature: Climate Change* 3 (2013), 975–78. doi:10.1038/nclimate1981.

Unschuld, Paul U. *Medicine in China: A History of Ideas.* Berkeley: University of California Press, 1985.

U.S. Census Bureau. "State & County QuickFacts." http://quickfacts.census.gov/qfd/states/00000.html.

U.S. Energy Information Administration. "International Energy Statistics." www.eia.gov/cfapps/ipdbproject/IEDIndex3.cfm?tid=90&pid=44&aid=8.

U.S. Environmental Protection Agency. "Sources of Greenhouse Emissions." www.epa.gov/climatechange/ghgemissions/sources.html.

U.S. Geological Survey, Northern Rocky Mountain Science Center. "Retreat of Glaciers in Glacier National Park." http://nrmsc.usgs.gov/research/glacier_retreat.htm.

Van Dam, Rob. "Ask the Expert: Coffee and Health." Harvard School of Public Health. www.hsph.harvard.edu/nutritionsource/coffee/.

Wen, Jian Min, and Garry Seifert, trans. *Wen Bing Xue: Warm Disease Theory.* Taos, NM: Paradigm, 2009.

Whiteman, Gail, Chris Hope, and Peter Wadhams. "Climate Science: Vast Costs of Arctic Change." *Nature* 499 (2013), 401–3. doi:10.1038/499401a.

Worsley, J. R. *Traditional Acupuncture: Traditional Diagnosis.* Taos, NM: Redwing, 1990.

Yale Environment 360. "Ocean Acidifying Faster Than Any Time in 300 Million Years." March 2, 2012. http://e360.yale.edu/digest/ocean_acidifying_faster_than _any_time_in_300_million_years_study_says/3357/.

Zhong-Jing, Zhang. *Shang Han Lun: On Cold Damage.* Trans. Craig Mitchell, Feng Ye, and Nigel Wiseman. Brookline, MA: Paradigm, 1999.

Index

A

Acupuncture, 47, 152, 153

Advanced Acupuncture (Cecil-Sterman), 152

Advertising. *See* Consumerism

Alcohol, 148, 149

American Cancer Society, 143, 145, 146, 158

Andes, 29–30

Anger, 100–101

Antarctica, 30

Arrhenius, Svante, 17

Assumptions, influence of, xi

Atlas of Chinese Tongue Diagnosis (Kirschbaum), 169

B

Balance
    creating personal, 46–48
    importance of, 41–44

Bian Que, 168

Bladder, 169, 170, 180–82

Blanc, Mont, 29

Brain, in Chinese medicine, 169

C

Canadell, Josep, 22

Cancer
    as big business, 157, 159
    climate change and, xiv, 156, 165
    coffee and, 149–51
    deaths from, 147
    deeper issues of, 151–54
    Eastern vs. Western approaches to, 147–51, 156–60
    prevalence of, 143, 146–47, 156
    responding to diagnosis of, 160–65
    roots of, 144–46
    secondary, 158
    statistics on, 143, 146–47
    treating, 152–54

Cecil-Sterman, Ann, 152

Centipede, 152–53, 162

Champlain, Lake, 15

Change
    as central tenet of Chinese medicine, 84
    climate change and, 139–40
    wind and, 87

Chemotherapy, 158

Chinese medicine. *See also* Five Phases
  tradition; Tongue diagnosis
    balance and, 47
    brain in, 169
    change in, 84
    climate change viewed through,
      xiii–xiv, xvii–xviii, 4–5, 12–13, 18,
      32–34, 92
    diagnostic methods of, 6–7
    fluids in, 155
    historical texts of, 8–9, 67
    holistic view of, xii–xiv, xvii, 7–9,
      34, 78
    inductive thinking and, 7–8, 32, 69, 78
    jing in, 71–72
    nutrition and, 49
    organs and, 9–10, 68
    symptoms and, xii–xiii, 3–5, 93
    treatment approaches in, 152–54
    water in, 23, 36, 197
Climate change
    anger and, 101–2
    assumptions and, xi–xii, 4
    basics of, 16–17
    cancer and, xiv, 156, 165
    coffee and, 60–65
    deforestation and, 18–22
    early warnings about, 13
    Eastern perspective on, xiii–xiv, xvii–
      xviii, 4–5, 12–13, 18, 32–34, 92
    effects of, 4, 13, 17–18, 92
    feedback loops and, 75
    global temperature increase limits
      and, 24
    ice sheets and glaciers and, 28–32
    as internal condition, 35–37
    ocean acidification and, 22–24
    as opportunity, 37, 192–98
    permafrost and, 25–28
    proportionate response to, 93–96
    scientific consensus on, 12, 17
    as symptom of deeper issues, xii, 4,
      13, 165
    treating underlying causes of, xiv, 5,
      14, 139–41, 195–98

    values and, 37–40
    wind and, 87–89
CLL (chronic lymphocytic leukemia),
    160–63
Coffee
    cancer and, 149–51
    climate change and, 60–65
    common attitudes toward, 57–60
    context of, 56–57
    decaf, 58
    eliminating, 60, 65
    fair-trade, 60
    iced, 58
    nature of, 50–54
    personal effects of, 49–50, 54–56,
      57–59, 195
    statistics on consumption of, 61–62
    substitutes for, 62–64
Cold, School of, 8, 9
Communication, 129–34, 140–41, 191
Conflict, 103–7
Connor, Steve, 29
Consumerism, 44–46, 121–22
Control cycle. *See* K'o cycle
Cuscuta, 77

D

Dandelion root, 62–64
Deforestation, 18–22, 33
Dimethyl sulfide (DMS), 24
Divine, experience of, 110, 111, 116–18, 119
*Divine Farmer's Materia Medica. See Sheng
  Nong Ben Cao Jing*
Dryness, concept of, 109
Du Qing-Bo, 168

E

Eaarth (McKibben), 17, 74
Earth phase, 189–91
Economics
    continuous growth and, 84–85,
      116–17
    energy and, 86

quality and quantity in, 121–24
Yin and Yang of, 89–92
Eight Extraordinary Meridians, 152
Elijah, Thea, 112
The End of Nature (McKibben), 13
Exercise, 154–55

F

Fear, 70–71, 143
Fire
    associations of, 126
    communication methods and,
        129–33, 191
    deficiency of, 128
    excess of, 128–31, 139–41, 144
    relationship of Water and, 136–38
    relationship of Wood and, 125, 135–36
    toxins, 171
    warmth of, 126–28
    Yin of, 133–34
Five Phases (Five Elements) tradition. *See
    also individual phases*
    balance and, 68
    conflict and, 103–5
    correspondences in, xiv, 68
    growth and, 90–91
    interpretations of, 68
    K'o cycle, xvi, 109, 115, 136
    seasons in, xiv, 194
    Sheng cycle, xvi, 82–84, 85–86, 109,
        119, 136
Flaws, Bob, 148, 149
Fluids, Eastern view of, 155
Forests. *See also* Deforestation
    cooling effect of, 22
    pine beetles and, 20–21
    planting, 20
    Yin nature of, 18–19, 36

G

Gallatin Valley, 20–21
Gallbladder
    overstimulation of, 100–101, 106

tongue diagnosis and, 169, 170, 172,
    178–79
Gandhi, Mahatma, 47
Ginger, 153
Glacier National Park, 31–32
Glaciers, 28–32, 33, 36
Global warming. *See* Climate change
Gore, Al, 12
Greenhouse gases. *See also* Climate change
    climate change and, 17
    deforestation and, 20, 21–22
    in permafrost, 27–28
    sources of, 37–38
Greenland, 29
Grief, 112–13
Growth
    belief in continuous, 84–86, 90, 115,
        116–17, 119, 159
    consequences of continuous, 84–85
    meaning and, 115–18

H

Hansen, James, 13, 23, 32, 75
Hanson, Greg, 2
Harvest season, 187–89
Health
    Five Phases and, 68
    Yin and Yang and, 42–43
Heart
    communication between Kidney
        and, 137
    Fire and, 126
    role of, 10, 126
    tongue diagnosis and, 169, 170,
        173–74
    Yin of, 134
Heat
    concept of, 10, 109
    creation of, 10
    from electronic devices, 130
    School of, 8–9
    Yin-deficient, 10, 13, 33, 77, 132
Heavenly Qi, 111, 112, 114
Himalayas, 29

I

Ice sheets, 28–32, 33, 36
Inductive thinking, 7–8, 32, 69
Intergovernmental Panel on Climate
    Change (IPCC), 12, 20, 24, 32
Irene, Tropical Storm, 1–3, 15, 69

J

Jing
    concept of, xv–xvi, 71–72
    as energetic reserve, 72
    herbs for tonifying, 77–78
    Kidneys and, 71, 81
    planet's, 72–77, 78
    Qi and, 72
    wisdom and, 71, 75–76

K

Kidney
    communication between Heart and,
        137
    fear and, 143
    importance of, 69–70
    jing and, 71, 81, 137
    tongue diagnosis and, 169, 170, 172,
        180–82
K'o (Control) cycle, xvi, 109, 115, 136
Kort, Eric, 26

L

Large Intestine
    Metal and, 110, 111
    role of, 111–12, 123
    tongue diagnosis and, 182–83
Licorice, 153
Liver
    overstimulation of, 100–101, 106
    role of, 9–10
    tongue diagnosis and, 169, 170, 172,
        178–79

Lung
    Metal and, 110, 111
    role of, 9, 111, 112, 123
    tongue diagnosis and, 169, 170,
        175–76

M

Maciocia, Giovani, 148
Marshall, Michael, 27
McKibben, Bill, 13, 17, 74
Meaning
    growth and, 115–18
    wisdom and, 118–19
Meat, 149
Meditation, 47
Meridians, 152, 197
Metal
    associations of, 110, 111–13
    experiencing, 113–14
    lack of, 144
    relationship of Water and, 118–19
    relationship of Wood and, 115–18,
        120–24
Methane, 25–27, 37
Method, as message, 100–103
More, limits of, 154–55

N

Nan Jing: Classic of Difficulties, 152
Nature, view of, 103–5
Nei Jing. See Yellow Emperor's Classic of
    Internal Medicine
Nourishing cycle. See Sheng cycle

O

Oceans
    acidification of, 22–24, 33
    cooling effect of, 23–24
    Yin nature of, 22–23, 36
Oil, 72–77, 78, 86

Organs
   in Chinese medicine, 9–10, 68
   in Western medicine, 143, 157

P

Pericardium
   Fire and, 126
   role of, 126
   tongue diagnosis and, 169, 170,
      173–74
Permafrost, 25–28, 33, 36
Phases. *See* Five Phases tradition
Phytoplankton, 24
Pierce, Fred, 17
Pine beetles, 20–21
Polygonum, 77
Proportionality, concept of, 94–95
Pulse diagnoses, 6, 9

Q

Qi
   coffee and, 50
   concept of, 6
   distribution of, 176
   exercise and, 154
   Five Phases and, 82
   green tea and, 62
   Heavenly, 111, 112, 114
   jing and, 72
   listening to pulses of, 6
   meridians and, 197
   overuse of, 10
   strengthening, 153
Qi Gong, 47, 72, 114, 118, 119
Quality and quantity, 121–24
Qu Ci Shan, 148

R

Rabatel, Antoine, 30
Radiation therapy, 158
Rehmannia, 77, 153, 162
Robbins, Jim, 21

S

School of Cold, 8, 9
School of Heat, 8–9
Scorpion, 152–53, 162
Scrophularia, 153, 162
Semiletov, Igor, 25–26
Separation, lens of, 3–4, 7, 157
Shang Han Lun, 8
Shen, 137
Sheng (Nourishing) cycle, xvi, 82–84,
   85–86, 109, 119, 136
*Sheng Nong Ben Cao Jing (Divine Farmer's
   Materia Medica),* 153
Small Intestine, 126, 182–83
Smoking, 148–49
Spleen, 169, 170, 176–78
Stomach, 169, 170, 176–78
Sulfur, atmospheric, 24
Sweating, 154–55
Symptoms
   diagnosing and treating root causes
      of, xii–xiii, 3–5
   Eastern vs. Western view of, 93
   interconnection of, xiii
   as messengers, xiii, 3, 93–94

T

Tai Chi, 47, 72, 113, 114, 119
Tea
   dandelion root, 62–64
   green, 62
Tobacco, 148–49
Tongue diagnosis
   back, 180–82
   behind the tip, 175–76
   coating, 171
   color, 170–71
   concept of, 6, 9, 168–69
   history of, 168
   map, 169–70
   middle and center, 176–78
   opportunity of, 183–86

Tongue diagnosis (*continued*)
  outside middle, 178–80
  sides of the back, 182–83
  size/shape, 172
  texture, 171
  tip, 173–74
Trees. *See also* Deforestation
  pine beetles and, 20–21
  planting, 20
  Yin nature of, 18–19, 36
Triple Heater, 126

U

Unschuld, Paul, 104

V

Values, importance of, 37–40
Vermont, extreme weather in, 1–3, 15–16,
  81–82

W

Warring States era, 104–5
Water
  associations with, 69–70
  in Chinese medicine, 23, 36
  fear and, 70–71
  lack of, 144
  relationship of Fire and, 136–38
  relationship of Metal and, 118–19
  relationship of Wood and, 81, 86, 88,
    100, 109, 115–16
  wisdom and, 71
  Yin and Yang of, 69–70
Wen Bing Xue, 8–9, 161
Wind
  change and, 87
  climate change and, 87–89
  in health and sickness, 87–89
  internal, 96–100, 109
Winooski River, 1–3
Wisdom, 71, 75–76, 118–19, 136–38
With Speed and Violence (Pierce), 17

Wood
  anger and, 100–101
  in balance, 87
  excess of, 85, 89–92, 105, 106–7, 110,
    135–36, 144, 146, 152, 156, 190, 192, 195
  relationship of Fire and, 125, 135–36
  relationship of Metal and, 115–18,
    120–24
  relationship of Water and, 81, 86, 88,
    100, 109, 115–16
  value of, 105
  wind and, 87
  Yin of, 106–7
Worsley, J. R., 68
Wu xing. See Five Phases tradition

Y

Yang
  balancing Yin and, 41–44, 47
  concept of, 10–11
  consumerism and, 44–46
  overemphasis on, 38–41, 84, 85,
    133–35, 146, 152, 192, 195
  of unfrozen states, 25
  of water, 69–70
*Yellow Emperor's Classic of Internal Medi-
  cine (Nei Jing)*, 8, 9, 67–68, 83, 168
Yin
  balancing Yang and, 41–44, 47
  of communication, 133–34
  concept of, 10–11
  -deficient heat, 10, 13, 33, 77, 132
  of frozen states, 25
  health and, 35–36
  of oceans, 22–23
  satisfaction and, 44
  of trees, 18–19
  undervaluation of, 39–40, 152, 192
  of water, 69–70
Yin-Yang symbol, 43, 91
Yuen, Jeffrey, 152

Z

Zhao Xue Min, 149